D1163686

Mercury's perihelion
FROM LE VERRIER TO EINSTEIN

Mercury's perihelion

FROM LE VERRIER TO EINSTEIN

N.T. ROSEVEARE

CLARENDON PRESS · OXFORD
1982

Oxford University Press, Walton Street, Oxford OX2 6DP
London Glasgow New York Toronto
Delhi Bombay Calcutta Madras Karachi
Kuala Lumpur Singapore Hong Kong Tokyo
Nairobi Dar es Salaam Cape Town
Melbourne Wellington
and associate companies in
Beirut Berlin Ibadan Mexico City

Published in the United States by
Oxford University Press, New York

British Library Cataloguing in Publication Data

Roseveare, N.T.
Mercury's perihelion from Le Verrier to Einstein.
1. Mercury (Planet)
I. Title
523.4'13 QB611

ISBN 0-19-858174-2

Filmset and printed in Great Britain
by the Alden Press, Oxford

ACKNOWLEDGEMENTS

This book is a substantial revision of my doctoral thesis 'A history of the anomalous advance in the perihelion of Mercury', accepted by the University of London in 1978. I would like to thank Professor H.R. Post and the rest of the Department of History and Philosophy of Science at Chelsea College, London for everything I learnt from them between 1974 and 1977. In particular my thanks are due to my supervisor Jon Dorling for his help and encouragement, not only while I was undertaking the research but also after it had been completed and written up. I am grateful to the Department of Education and Science for the award of a Major State Studentship held from 1974 to 1977.

A number of people have made helpful comments on the text; amongst them I thank Professor I.W. Roxburgh, Dr N.J. O'Riordan and Mrs J.A. Redfearn. I would also like to thank Mrs P. Jessop for translating part of a paper by Buys-Ballot and Mr A.T. Sinclair of the Royal Greenwich Observatory for help with a problem in celestial mechanics.

Most of the research was carried out in the Science Museum Library in London and I wish to express my thanks for the help given and courtesy shown to me by the staff of the Library, who made it a pleasure to study there. I also spent some time at the library of the Royal Astronomical Society, and I wish to thank both the Council of the Society for permission to use their library and the Librarian Mrs Enid Lake for help and encouragement.

I would also like to thank my parents for their invaluable support, interest, and encouragement during my time at university, and my wife Tessa for help with the final preparation of the original thesis, with the preparation of the typescript of the book, and for support and encouragement throughout.

London N.T.R.

CONTENTS

END

1

INTRODUCTION

1.1. The Mercury anomaly

The planet Mercury is both the smallest major planet in the solar system and the nearest to the Sun. These circumstances have always combined to make it a difficult planet for astronomers to observe. There is a tradition that Copernicus lamented that he had never seen it. Astronomers of the seventeenth and eighteenth centuries found that their observations of Mercury were of insufficient accuracy to give a satisfactory theory of its motion. By the beginning of the nineteenth century, when a good theory should have been achieved, this small planet still failed to move according to predictions. The French astronomer Urbain Jean Joseph Le Verrier started work on Mercury's motion early in his career. Born in 1811, Le Verrier became director of the Paris Observatory in 1854. It was a post which ranked as equivalent to the English Astronomer Royal, and involved similar administrative duties. Le Verrier proved to be unpopular in carrying out these duties and was replaced as director by another mathematical astronomer, Charles Delaunay. Le Verrier's scientific career flourished unaffected by this; he eventually regained the post in 1873 after Delaunay's death and held it until he himself died in 1877.

Le Verrier produced a theory of Mercury's motion in 1843, but it was not perfect. In reworking the theory with the most modern and accurate observations available, he found what he thought to be a basic problem; in 1859 he announced that he had discovered an anomalous advance in the perihelion of Mercury which was producing the discrepancy between the predictions and observations of its motion. The nature of this effect will be explained in the next section. Le Verrier was proved correct in his analysis of Mercury's motion, but his proposed solution that planetary matter would be discovered between the Sun and Mercury was not verified. In 1882 the leading astronomer and superintendant of the American nautical almanac, Simon Newcomb, verified Le Verrier's theory of Mercury's motion. In 1895 Newcomb produced a detailed work concerning the orbits of Mercury, Venus, Earth, and Mars which became the basis for the tables of these planets' motions in the main national ephemerides. His solution to the perihelion advance was an alteration to the Newtonian law of gravitation, a change which made the tables correct but which had to be abandoned within a few years for other reasons. One of Germany's top astronomers, Hugo von Seeliger, director of the Munich Observatory from 1882 to 1924, proposed the next important hypothesis. Seeliger suggested that matter surrounding the Sun caused the perihelion advance of Mercury. It was a hypothesis similar to Le Verrier's, but it had observational support in the form of the zodiacal light, and it found general acceptance.

At the same time as these and numerous other astronomers were working on the Mercury problem, a number of physicists had been investigating Newton's law of gravitation to see if alterations to that law might explain the perihelion advance. Physicists from Wilhelm Weber, the nineteenth century Göttingen master of electrodynamics, to Walther Ritz, the young Swiss who died in 1909, became involved. Scientists of note and others quite obscure made their contributions. As the twentieth century began, a new physics arose. Relativity became of fundamental importance, with Einstein's special theory taking the central position. The new physicists raced to produce a theory of gravitation which satisfied both their own theoretical tenets and also the demands of observation, until at the end of 1915 Einstein published his general theory of relativity. As a consequence of the theory, he showed with triumph that he could account for all the anomalous perihelion advance that Le Verrier and Newcomb had found. The gradual acceptance of general relativity led to the abandoning of Seeliger's zodiacal light hypothesis, though rearguard actions were fought on scientific and unscientific grounds for some time. As late as the 1960s and 1970s attempts were made to use the Mercury advance to distinguish between general relativity and the rival scalar–tensor theory of R.H. Dicke.

The success of Einstein's theory has ensured the fame of the Mercury advance, while at the same time obscuring the complex history of the attempts made to solve the problem. The purpose of this book is to counter the simple accounts and to present the detailed history of these attempts and to place them in their scientific context. In addition an account of the development of the new gravitation theories is given which has no claims to being a full history, but which presents them purely from the aspect of their effect on planetary perihelion motions, in particular the anomalous part of Mercury's motion. From Le Verrier to Einstein a wealth of science was produced in response to the anomalous advance in the perihelion of Mercury.

1.2. The nature of anomalies

Le Verrier had found that the errors in his predictions of Mercury's positions could be traced back to the behaviour of its perihelion, which is the point on the elliptical orbit of a planet nearest the Sun. Due to the gravitational influence of other planets on Mercury this point is not fixed but moves round in the direction of Mercury's motion; the perihelion is said to advance. Le

Verrier discovered that the advance in the perihelion of Mercury was greater than could be accounted for by adding together the gravitational effects of the known planets. It was the discrepancy between the observed value and the theoretical value predicted from Newtonian theory that caused the advance to be anomalous. The magnitude of the discrepancy is the anomalous part of the advance, often just known as the anomalous advance.

The word 'anomaly' has become a technical term in the history of science, denoting an empirical observation which disagrees with a theoretical prediction (i.e. an observation is an anomaly for a particular theory). The concept of anomaly is thought to play an important role in the progress of scientific theories. Serious anomalies, which cannot be explained as observational error or by the omission of information, may lead to the overthrow of a theory, especially if there is a rival theory giving good agreement with experiment. The former theory is then said to have been 'refuted' by the anomaly. The philosopher Sir Karl Popper attributes much of the progress of scientific knowledge to this process.

Since predictions change as theories develop, and observations improve with experimental techniques, so disparities between predictions and experiences also change. In some cases, however, old theories are defended by the introduction of additional hypotheses, or the anomaly may be stubbornly ignored. The fact that some theories are not discarded even in the face of disagreement with experience led Imre Lakatos to propose a refinement of Popper's ideas, namely his methodology of research programmes, in which theories contain 'hard cores' which are immune to refutation.

The philosophical importance of the anomaly to the progress of scientific theories justifies devoting an entire book to the subject rather than to the work of a particular scientist, the development of a new concept, or a new discovery. This book covers almost the entire range of options open to scientists faced with an anomaly to a theory. It discusses the options chosen by the conservatives, who tried to contain the anomaly within the conventional theory (from Le Verrier right up to Charles Lane Poor in the 1920s); by the radicals who tried to alter the conventional theory (from Challis in 1859 through Einstein up to Dicke in the 1960s); by those who try to bypass the problem as best they can (Simon Newcomb at times); and by those who say little (such as John Couch Adams).

At the turn of the century, the problem with Mercury's orbit was one of the main sources of concern for those astronomers interested in the motions of solar system bodies. Two other problems also occupied them: the irregular acceleration of Encke's comet and the secular acceleration in the Moon's mean motion. The acceleration of Encke's comet had been explained by J.F. Encke soon after the comet had been discovered in 1818, as resulting from a resisting medium surrounding the Sun. But that idea had to be abandoned when the acceleration was later found to be irregular. Recent explanations emphasize the role of the physical constitution of the comet, as F.D. Whipple's icy-conglomerate model of the comet nucleus. The secular acceleration of the Moon's mean motion had been a well-known problem since Edmund Halley had found from ancient eclipse observations that the Moon appeared to be slightly but perceptibly accelerating. It was thought in the late eighteenth century that Laplace had solved the problem using Newtonian perturbation theory, but in 1853 and 1859 John Couch Adams showed that gravitational action could account for only half the observed value of the acceleration. The remainder was then generally considered as being due to tidal friction, an effect put into quantitative terms in the 1920s but the adequacy of which has been recently questioned (see Munk and MacDonald 1960).

Neither the problem of Encke's comet nor that of the lunar acceleration was considered an anomaly for the Newtonian inverse square law of gravitation, since alteration of that law would not readily have given the required motions. They were considered as serious anomalies within the wider gravitational theory which included not only the central force law but additional laws and hypotheses governing the application of that force law to the actual conditions pertaining in the solar system. The anomaly of Mercury's perihelion was problematic because it was difficult to decide whether it arose from the wider theory or from an error in the inverse square law of gravitation itself. In the 10th edition of the *Encyclopaedia Brittanica*, Simon Newcomb cited the perihelion advance of Mercury and the secular acceleration of the Moon's mean motion, in that order, as the two most important anomalies in theories governing the motion of the solar system (Newcomb 1902). J. Zenneck, in a 1903 review of gravitation, cited the Mercury discrepancy (and three lesser problems that Newcomb had discovered), Encke's comet and the lunar secular acceleration as the greatest problems in gravitation (Zenneck 1903). Henri Poincaré, in lectures delivered in 1906, also listed all three problems as significant for gravitational theory (Poincaré 1953).

In 1915, Einstein was able to give a clear account of Mercury's perihelion advance directly from his theory of gravitation. As soon as it was accepted that the advance was a relativistic effect, the inverse square law was identified as the culprit for Le Verrier's shifting perihelion. In relating the history of the proposed solutions from Le Verrier to beyond Einstein, this book sets out to illuminate the complexities facing Le Verrier's successors. It is only with hindsight that the problem appears clear cut; the choices available to those astronomers faced with the anomaly were by no means easy to resolve. In order to appreciate their problems clearly, it is necessary to introduce the basic properties and notations of elliptical orbits, and to show the fundamental effects of the inverse square law of gravitation and laws closely approximating to it. Finally accounts must be given of the state of knowledge concerning the solar system in the middle of the nineteenth century and of the general regard with which celestial mechanics and its results were held at that time.

1.3. Elliptical orbits

Kepler's first law states that a planet moves in an elliptical orbit which has the Sun at one focus (see Fig. 1.1).

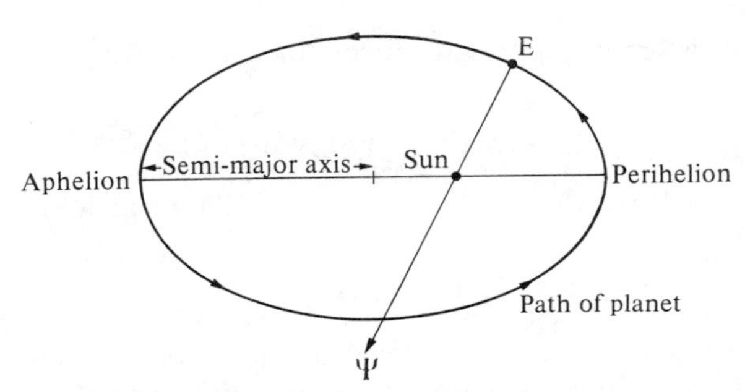

FIG. 1.1 Plan of an elliptical orbit.

The point on the ellipse nearest the Sun is called the perihelion, and the point furthest away is called the aphelion. For the Moon's orbit round the earth the corresponding points are called the perigee and apogee respectively. If the eccentricity of a planet's orbit is e, and the semi-major axis is a, then the distance between the sun and the perihelion is $a(1-e)$. The plane in which the Earth's orbit lies is known as the plane of the ecliptic, and this is used as the base from which the inclinations of the orbits of the other planets are measured. The Earth passes through perihelion on or about 2 January. When the earth is at point E in Fig. 1.1 it is at the vernal equinox, and both day and night last 12 hours. At point E, astronomers on the Earth observe the Sun to lie in the direction of Ψ, the first point of Aries. Ψ is chosen as a fixed point from which planetary longitudes are measured.

Figure 1.2 shows the orbit of a second planet inclined to the ecliptic at an

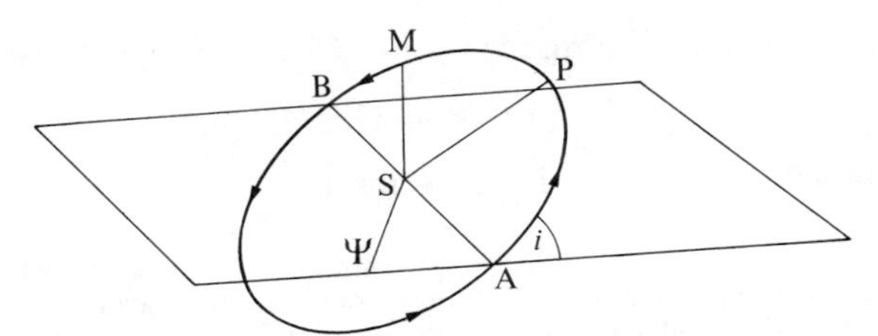

FIG. 1.2 An inclined elliptical orbit.

angle of i. The two orbital planes cut along the line AB, called the line of nodes; A and B are known as the ascending node and descending node respectively. P is the perihelion of the planet, and by convention the longitude of perihelion ϖ(curly pi) is measured from Υ. The angle $\theta = <\Upsilon SA$ is the longitude of the ascending node A, and the angle $\omega = <ASP$ is the angle from the node to the perihelion. Adding these gives the longitude of perihelion, so that

$$\varpi = \theta + \omega. \tag{1.1}$$

If the planet is at any point M in its orbit, then its true longitude is said to be

$$L = \theta + \omega + v$$

or

$$L = \varpi + v \quad \text{using eqn (1.1).} \tag{1.2}$$

Where $v = <PSM$, known as the true anomaly.† The mean longitude l at time t is given as

$$l = \varpi + n(t - \tau). \tag{1.3}$$

In eqn (1.3), n is the mean angular velocity or mean motion, defined in terms of the period of revolution of the orbit, i.e. $n = 2\pi/T$, where T is the period. Putting $t = 0$ gives

$$l = \varpi - n\tau \tag{1.4}$$

which defines a longitude $l = \varepsilon$ known as the longitude of the epoch. Putting $t = \tau$ in eqn (1.3) gives $l = \varpi$ so that the planet is at perihelion at this time. The expression $n(t - \tau)$ is defined to be the mean anomaly M, so that eqn (1.3) may be written

$$l = \varpi + M. \tag{1.5}$$

The mean anomaly M, the eccentricity e and the true anomaly v are connected by the 'equation of the centre'

$$v - M = (2e - \tfrac{1}{4}e^3)\sin M + (5/4)e^2 \sin 2M + (13/12)e^3 \sin 3M + \ldots \tag{1.6}$$

A proof of this may be found in Smart (1965 p. 119).

To first order in e this is

$$v - M = 2e \sin M. \tag{1.7}$$

Rearranging and substituting for v and M from eqns (1.5) and (1.2) gives

$$L = l + 2e \sin(1 - \varpi). \tag{1.8}$$

This equation will be used in a later section.

† The word 'anomaly' has a technical meaning in astronomy which must not be confused with the meaning it has in the philosophy of science. In describing an elliptical orbit, the anomaly is the angular distance through which the planet has moved since perihelion, e.g. true anomaly, mean anomaly, and eccentric anomaly.

The orbits that have been briefly described here give the positions of planets with respect to the Sun. Heliocentric coordinates must be changed into geocentric coordinates for the use of terrestrial observers. The key factor relating the two types of coordinate is the position of the Sun, and an accurate theory of the Sun's motion with respect to the Earth is needed on which to base the theory of a particular planet's heliocentric motion. In addition the astronomer must take into account the effects of precession, nutation, aberration, refraction, and parallax (the natures of which need not concern us here) in order to obtain results applicable to the actual conditions pertaining in the solar system. Finally corrections need to be made for instrumental errors and any errors peculiar to the method of observation.

When precise positions of heavenly bodies are required meridian observations are usually made. For Mercury and Venus transit observations are also available, as their orbits lie between that of the Earth and the Sun. Figure 1.3 shows the orbits of Mercury (M) and the Earth (E) around the Sun, with the earth fixed for example at E. At the point M_1, Mercury is nearest to the Earth and is said to be at inferior conjunction. It is at superior conjunction when it is furthest from the Earth, at M_2. The furthest angular distance from the Sun that Mercury can appear to have is 28°, when it is at M_3 and again at M_4.

Meridian observations are made when the planet crosses the observer's meridian, which is an imaginary great circle across the sky along a north–south axis and passing through a point vertically above the observer. Since Mercury is never more than 28° from the Sun, it crosses the observer's meridian within two hours of the Sun's passage at noon. Meridian observations of Mercury therefore take place in daylight. Transit observations also take place in daylight, since a transit of Mercury occurs when it appears to

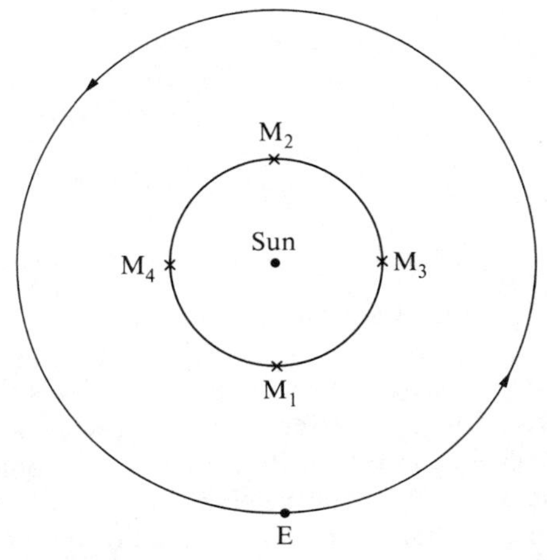

FIG. 1.3 Important points in the orbit of Mercury.

pass in front of the Sun. The silhouette of the planet may be viewed moving across the solar disc. Transits also occur for Venus, but not for planets whose orbits lie outside that of the Earth. Figure 1.3 shows that a transit can only occur when the inner planet is at inferior conjunction, but not at every inferior conjunction does a transit occur, as Fig. 1.4 shows.

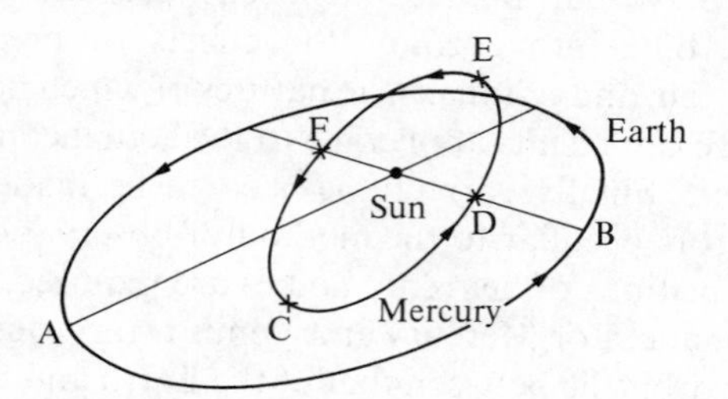

FIG. 1.4 Orbital conditions for the occurrence of a transit.

If the Earth were at A, Mercury would be at inferior conjunction at C and superior conjunction at E. A transit would not occur because Mercury passes 'below' the Sun at inferior conjunction. If the Earth were at B, a transit could occur, because Mercury at inferior conjunction is at D, which is directly in front of the Sun. As the Earth's orbit defines the ecliptic, and the points at which Mercury's orbit goes through the ecliptic are its nodes, transits will occur when Mercury passes through its ascending node (in November) or its descending node (in May) if the Earth is at about the same longitude as the node. The path of Mercury across the Sun's disc is shown in Fig. 1.5, in which Mercury's size has been exaggerated.

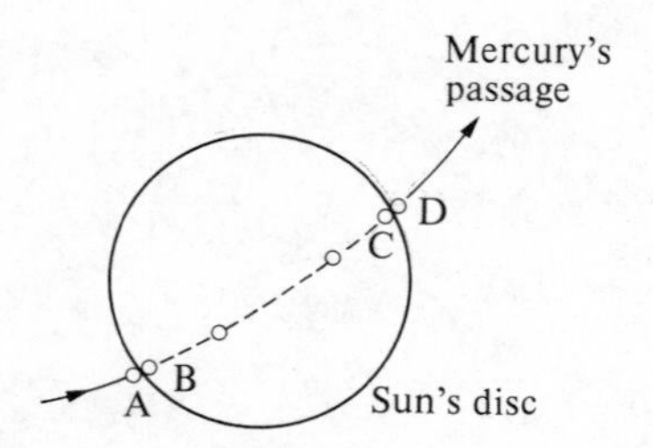

FIG. 1.5 A transit of Mercury.

Distinction is made between positions A and D and positions B and C. At A and D the planet is said to be in external contact, while at B and C internal contact is taking place. When the transits of Venus were found to be important for measuring the solar parallax† in the eighteenth and nineteenth centuries, much work was done to make the observations of contacts as precise and

† The solar parallax is a quantity describing the distance between the Earth and the Sun.

reliable as possible. Internal contacts were generally favoured as giving the better accuracy.

As transits do not take place at every inferior conjunction they are not common events. Future transits of Mercury occur in 1986, 1993, and 1999 while those of Venus are more rare, the next two occurring in 2004 and 2012, the last two having occurred in 1874 and 1882.

1.4. The inverse square law

The Newtonian theory of gravitation rests on the inverse square law of gravitation

$$\mathbf{F} = -G\frac{Mm}{r^2}\hat{\mathbf{r}} \tag{1.9}$$

where M is the mass of the central body, G is the universal constant of gravitation and $\mathbf{F}$ is the gravitational force on the body of mass m at a distance r from the central body. The unit vector $\hat{\mathbf{r}}$ is in the direction of the body of mass m and away from the central body, so that the negative sign indicates that the gravitational force acts towards the central body and is an attractive force. From this law each of Kepler's three laws may be derived—that planets move in elliptical orbits with the Sun at one focus; that equal areas are swept out by a planet in equal times; and that the cube of the period of an orbit is proportional to the square of the distance to the Sun (the semi-major axis).

In agreement with Newton's third law, which states that action and reaction are equal, there is an equal and opposite force on the Sun due to the Earth's attraction. The effect on the Sun's motion is much less than that on the Earth's motion, and the two move around their common centre of gravity near the centre of the Sun.

When accounting for the orbit of a planet, the effect of the other planets must be considered as well. Every particle in the solar system attracts every other particle, and though the major force in each case is that directed toward the central body (whether this is the Sun for the planets or a particular planet for its satellites), the other forces disturb or 'perturb' the standard elliptical orbit. The innermost planet in the solar system is Mercury, and its nearest neighbour is Venus. Let $\mathbf{S}$ be the gravitational force on Mercury due to the Sun, and let $\mathbf{P}$ be the force due to Venus. The angle between $\mathbf{S}$ and $\mathbf{P}$ depends on the configuration of Mercury and Venus. The total force on Mercury is now

$$\mathbf{F} = \mathbf{S}+\mathbf{P}$$

or

$$\mathbf{F} = -G\frac{Mm}{r^2}\hat{\mathbf{r}}+\mathbf{P}. \tag{1.10}$$

$\mathbf{P}$ is the perturbing force, and since the mass of Venus is much smaller than that of the Sun, **P** is of small magnitude compared to **S**. Terms corresponding to the perturbing forces of the other planets must also be added to eqn (1.10). Whereas Newtonian theory allows an exact solution to eqn (1.9)—the two body problem giving an elliptical orbit—it does not give such a solution when there are three or more bodies. Perturbation theory has developed techniques involving series solutions to cope with this, and can give approximate solutions to high degrees of accuracy. This of course requires a large number of terms—a lengthy business in the last century when a 'computer' was an assistant who carried out numerical calculations. Decisions on which terms to leave out of the solution sometimes led to incorrect answers, and a number of anomalies were resolved when terms thought to have no effect were included.

Whereas eqn (1.9) gives a closed elliptical orbit, eqn (1.10) gives an ellipse that does not close up. Instead of the angle between successive perihelia being $360°$, the angle becomes $360° + \mathrm{d}\varpi$, where $\mathrm{d}\varpi$ is small compared to $360°$. In this case the perihelion is said to advance an amount $\mathrm{d}\varpi$ during each revolution, and the orbit precesses. Figure 1.6 shows the 'rosette' type of orbit resulting from an exaggerated precessing ellipse.

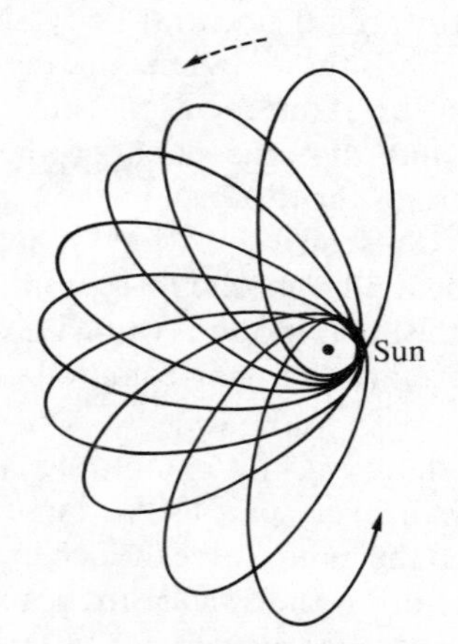

FIG. 1.6 A precessing elliptical orbit.

In the case of Mercury, planetary perturbations produce an advance of perihelion of just over $500''$ per century, which as Mercury undergoes about 4 orbits per year, is equivalent to an angle of $1\frac{1}{4}''$ per orbit. The ellipse would take about 260 000 years to precess through one full revolution.

The planetary composition of the perturbations for Mercury's perihelion, as given by Le Verrier in 1859, are given in Table 1.1. Mercury's neighbour Venus is the dominant planet but the giant Jupiter still manages to contribute just over a quarter of the total.

Table 1.1

Venus	280″.6
Earth	83″.6
Mars	2″.6
Jupiter	152″.6
Saturn	7″.2
Uranus	0″.1
Total per century:	526″.7

(Le Verrier 1859 p. 99).

Any such perihelion advance has to be taken into account when the orbit of a planet is discussed. From eqn (1.8) we have for the true longitude:

$$L = l + 2e \sin (l - \varpi)$$

Differentiating with respect to time gives:

$$\frac{\mathrm{d}L}{\mathrm{d}t} = 2\frac{\mathrm{d}e}{\mathrm{d}t} \sin (l-\varpi) - 2e \cos(L-\varpi)\frac{\mathrm{d}\varpi}{\mathrm{d}t}. \qquad (1.11)$$

If a planet has a small eccentricity the perturbation of the perihelion would have little effect on the planet's longitude since $\mathrm{d}\varpi/\mathrm{d}t$ is multiplied by e. Mercury had the largest eccentricity (0.206) of the major planets known in 1859. If this had not been the case Le Verrier might not have discovered the anomalous value for $\mathrm{d}\varpi/\mathrm{d}t$.

Substituting standard values for e and ϖ of Mercury and the required value of l, we obtain an equation of the form;

$$\frac{\mathrm{d}L}{\mathrm{d}t} = k_1\frac{\mathrm{d}e}{\mathrm{d}t} - k_2\frac{\mathrm{d}\varpi}{\mathrm{d}t}, \qquad (1.12)$$

in which k_1 and k_2 are real constants. It can now be seen directly how the values of $\mathrm{d}e/\mathrm{d}t$ and $\mathrm{d}\varpi/\mathrm{d}t$ affect the longitude of the planet. Le Verrier in 1859 gave $\mathrm{d}\varpi/\mathrm{d}t = 526''.7$ and $\mathrm{d}e/\mathrm{d}t = 4''.2$ for the total planetary perturbations per century. The corresponding value of $\mathrm{d}L/\mathrm{d}t$ must be taken into account when the longitude of Mercury is being computed. If a mistake were made, in the mass of Venus for example, there would be an error in $\mathrm{d}L/\mathrm{d}t$ and hence in the value for L, the true longitude of Mercury. The observed position would not then agree with the position predicted from theory.

If the reverse happened, and a disagreement with theory occurred when it was thought to be correct, then the theory would have to be checked. It might be that a planetary mass was in error, though since each planet perturbs a number of others its mass could not be arbitrarily changed if its effect on those other planets already produced agreement with observations of them. The relationships between planetary masses and orbital elements predicted from perturbation theory were fully known in 1859. Nevertheless as far as the perihelion was concerned, a completely different possibility was available.

The advancing perihelion has resulted from the introduction of the small term **P** in eqn (1.10). If the fundamental gravitational law were not that of eqn (1.9) but involved a second term such as in eqn (1.10), depending perhaps on the inverse cube of the distance, then an advancing perihelion would occur even in the two body problem. For the three body case the total perihelion advance would then consist of a part due to the gravitational law and a part due to perturbations. Any discrepancy in the observed advance could be ascribed either to the second term in the gravitational law or to new perturbations, and extra criteria would have to be brought in to decide between these two possibilities.

For a force law of the general form (in scalar notation)

$$F = \frac{G\,Mm}{r^n}, \tag{1.13}$$

the angle between successive perihelion may be shown to be

$$\mathrm{d}\varpi = 2\pi(3-n)^{-\frac{1}{2}} \tag{1.14}$$

For the inverse square law we have $n=2$, or $\mathrm{d}\varpi=2\pi$, so that the planet moves through 2π between perihelia and the perihelion itself does not move. For $n=2+\delta$, where δ is a very small number, the angle between successive perihelia is $2\pi(1-\delta)^{-\frac{1}{2}}$, or $2\pi(1+\frac{1}{2}\delta)$ to first order, and the perihelion advances an angle of $\delta\pi$ per revolution. A force law with such a distance dependence would approximate closely to the inverse square law but would produce a small advance of perihelion. Such a force law, with *ad hoc* δ to account for the observed discrepancy in Mercury's motion, was used in 1895 by Simon Newcomb to account for the anomaly. It will be discussed in detail in Chapter 5.

That laws differing from the inverse square law gave precessing orbits was known long before Mercury's anomalous advance was discovered. Such results may be found in Newton's *Principia*, the principle work in which Newton put forward his gravitational theory. Section IX of Book I of the *Principia* is called 'The motion of bodies in movable orbits; and the motion of the apsides'. Newton described what we know as the aphelion and perihelion as the upper apse and lower apse respectively, and the line between them as the line of apsides. He called the distance of the orbiting body from the central body the altitude, so that the greatest altitude occurred at the upper apse, and this greatest altitude Newton expressed 'arithmetically by unity' (Newton (1934) p. 144). Toward the end of Section IX the following proposition is introduced—'. . . if the centripetal force be as

$$\frac{b\,A^m - c\,A^n}{A^3},$$

the angle between the apsides will be found equal to

$$180^\circ \sqrt{\left(\frac{b-c}{mb-nc}\right)}, \qquad \text{(Newton p. 144).}$$

Here A represents the altitude or distance, b, c, m, and n are constants, and the angle considered is from one apse to the next. A single term force would be given by putting $c=0$, such that

$$F = b \cdot A^{m-3} \tag{1.15}$$

Newton's proposition tells us that the angle between successive perihelia will be

$$\begin{aligned} \mathrm{d}\varpi &= 2 \times 180^\circ \times \sqrt{\left(\frac{b}{mb}\right)} \\ &= 2\pi m^{-\frac{1}{2}}. \end{aligned} \tag{1.16}$$

This expression is identical to that given in eqn (1.14) for $m=3-n$, which would reduce eqn (1.15) to the form of eqn (1.13).

Newton was concerned with the theoretical effects of adding forces to the basic inverse square force in order to bring into his scheme the perturbing forces that inevitably result from universal gravitation, and which are needed to explain the departures from ordinary elliptical orbits observed for the planets and satellites. However Newton's theorems are equally valid if the two-term forces or the single-term forces departing slightly from the inverse square law form the basic law of gravitation itself. If the basic law did not provide a gravitational force with an exact inverse square dependence on the distance then the total theoretical perihelion advance would be greater than that obtained from the Newtonian law. In 1859 Le Verrier found that the observed perihelion advance of Mercury was about 39″ greater than the theoretical advance of 527″.

The extra amount required could have been ascribed to either of two quite distinct causes or some mixture of both. Firstly the anomalous advance could have been taken as being due to the perturbing effect of matter that had not been taken into account in the original theory. Such matter might have been a planetary mass which had been underestimated, or it might have been matter in the solar system that had not yet been discovered and whose influence had not been taken into account. If either case was accepted the Mercury advance would have been defused of its anomalous status and incorporated into Newtonian theory. Secondly the anomalous advance could have been taken as indicating that the Newtonian law was not exact, and that a new law had to be adopted which approximated to the old one in giving the confirmed motions of the solar system as well as this new perihelion advance. In this case the Mercury motion would have remained an anomaly for Newtonian theory, and indeed a highly important anomaly in causing this theory to be replaced by a new one.

Of course either of these two options had to be tested in relation to other effects. A planetary mass altered to produce Mercury's advance might produce discrepancies in other motions that had agreed with theory. A completely new body introduced as source of the perturbations might introduce other undesirable effects or it might not be observed. On the other hand a new law would have to have some sort of theoretical backing and again would have to be such that no new undesirable consequences were brought in.

A third strategy, that of ignoring the perihelion advance as an anomaly, would hardly have appealed to the mathematical astronomers of the time who had become used to explaining every detailed motion in the solar system. A fourth possibility was also available—that a mistake had been made in the theoretical work and no such anomalous motion really occurred. But this was difficult to test because the advanced nature of the work meant few people could tackle it and it would take a long time if they did. These two possibilities were not influential, and the main story of Mercury's perihelion splits into two parts. Either: (i) there was additional perturbing matter of some sort near the centre of the solar system; or (ii) the gravitational law operating was not exactly an inverse square law though this was for most purposes a satisfactory approximation. It was along these two lines that the debates, which started in the mid-nineteenth century and continued well into the twentieth century, ran their course.

1.5. The solar system in the mid-nineteenth century

In terms of the kinds of objects in it, the solar system known to Le Verrier was not greatly different from that known today. He knew the planets and their satellites, the minor planets or asteroids, the comets of long period and short period, and also the meteors. The number of known objects of course was smaller than now; the planet Pluto, for instance, was discovered in 1930. In addition our knowledge of the physical constitution of the solar system bodies has vastly improved, especially in the recent years of interplanetary space flight.

Ever since the telescope was invented there has been a steady discovery of new objects in the solar system. Most of these were due to chance and the combination of patient observing with astute identification, but a number were due to a deliberate prediction. In the eighteenth century a law was proposed which indicated that astronomers could be guided into discovering a new planet, one which would fill an apparent gap in the solar system. The law was called Bode's law after J.E. Bode, who did not state it first but who was responsible for making it well known. It was first written down by J.D. Titius in 1772 and quantified even earlier feelings about gaps in the solar system. Bode's law depends on the following series

$$0, 3, 6, 12, 24, 48, 96, \ldots$$

Adding 4 to each term gives

$$4, 7, 10, 16, 28, 52, 100, \ldots$$

These numbers correspond remarkably well to the mean distances of the planets from the sun. The mean distance of the Earth from the Sun is defined to be 1 astronomical unit or 1 au, and we have to multiply distances in au by 10 to obtain Bode's units. The actual mean distances of the planets known in 1772 were, in Bode's units,

Mercury (3.8), Venus (7.2), Earth (10.0), Mars (15.2),——, Jupiter (52.0), Saturn (95.5).

In 1781 William Herschel discovered what he thought to be a comet, but it was soon shown to be a planet with a mean distance of 19.2 au. When Bode realized how closely this agreed with the next term in his series, which was 192+4 or 196, he decided that the relationship had to be taken seriously. In particular he considered the apparent gap between Mars and Jupiter, and became convinced that a planet with a mean distance of 2.8 au was waiting to be discovered. He supported a mass search for the planet, but in January 1801, before this started, the Sicilian astronomer Giuseppe Piazzi discovered a planet-like object. Though the planet was then lost, Karl Friedrich Gauss developed new mathematical methods in order to compute an orbit from the little observational data available, and it was rediscovered at the end of 1801. The new planet Ceres had a mean distance of 2.767 au, confirming Bode's law, but its diameter was too small for a major planet. Soon a number of other asteroids or minor planets were discovered with similar mean distances from the Sun. Heinrich Olbers suggested that the original major planet between Mars and Jupiter had disintegrated at some time, and that Ceres and the other asteroids were fragments remaining from the explosion. Olbers's hypothesis implied that there were more asteroids to be discovered, and by 1850 a total of 12 had been found.

Bode's law was also used in the next discovery of planetary matter. The law indicated that any planet outside the orbit of Uranus would have a mean distance of 38.8 au (as 384+4=388). When in the 1840s Le Verrier and John Couch Adams started to try and find the orbit of the planet presumed by then to be perturbing Uranus, they both used Bode's law for their initial estimates of the new planet's mean distance. Both were successful, and consequently Neptune was discovered in 1846. Supporters of Bode's law were disappointed to find that the distance of the new planet was only about 30 au; much later the law was further discredited when Pluto was discovered with a mean distance of 40 au.

Astronomical discovery was also filling the solar system with objects other than planets and asteroids. Comets had always been discovered regularly, and in 1818 a particularly important one was observed by Pons. This was Encke's comet, named after the man who did so much theoretical work concerning it. Encke's comet was important for three reasons. First of all it was the first comet since Halley's comet which had been discovered and then had a return successfully predicted. The second reason was that its period was computed to be about $3\frac{1}{4}$ years, and thus was the first of the comets of short period. Finally Encke found that the comet was accelerating faster than gravitational perturbations could explain. He ascribed the anomalous acceleration to a resisting medium surrounding the sun, consisting of a tenuous fluid with a density rapidly increasing near its centre. The hypothesis survived until the late nineteenth century, when Encke's comet suffered variations in its acceleration which could not be explained with the continuous medium. Also certain comets had been found to be decelerating.

The solar system as known in 1850 was a busier place than that known in 1750. There were two more major planets, 13 minor planets, a debatable number of new satellites, new comets and meteors increasingly associated with periodic comets, as well as Encke's tenuous medium around the sun. Finally Laplace had assured the world that the solar system was stable and had provided, in his nebular hypothesis, a scientific description of its creation.

1.6. Celestial mechanics up to 1850

By 1850, just before Le Verrier discovered the Mercury anomaly, mathematical astronomers were optimistic. Using their analytical skills with Newtonian theory, they were by then able to explain every motion in the solar system that they were expected to explain. It was not a victory for mathematics alone, for the observational astronomers could claim their share; though, as has remained the case up to today, the theoreticians were accorded the major part of the glory. At the beginning of the seventeenth century Kepler had used the observations of Mars carried out by Tycho Brahe, who had been the leading astronomical observer in the world, to show that Mars had an elliptical orbit. The Newtonian principle of universal gravitation, introduced near the end of the seventeenth century, required that planets did not have stationary elliptical orbits, but ones distorted by neighbouring planets. As both observational techniques and theoretical treatment of observations improved, orbital deviations could be accurately measured. In turn, the mathematicians were able to describe the deviations accurately from theory. Sometimes the observers led the theoreticians and sometimes the reverse was the case, until, by about 1850, the motions of the solar system were thought to be completely explained by the Newtonian inverse square law of gravitation.

It had sometimes been a hard fight. The early Newtonians had to prevail against the Cartesians, who explained the motions of the planets using the vortex theory of Descartes. Even some of the early Newtonians were at times tempted to abandon the inverse square law for a near alternative. Clairaut, in company with d'Alembert and Euler, found in 1745 that he could not obtain the exact motion of the lunar perigee using the inverse square law.† He adopted a law with a distance dependence $1/r^2+a/r^4$, where a was a small constant, which, as can be seen from the previous section, gives a motion to the perigee. However when Clairaut took the inverse square law solution to a higher order of approximation he found that the correct lunar perigee motion was in fact obtained, and the Newtonian scheme remained intact.

Another problem in which choice of approximation played a part was tackled by Euler and Lagrange and eventually resolved by Laplace in 1773 and 1784. He showed that certain longstanding anomalies in the motions of Jupiter and Saturn were due to mutual perturbations of long period. Long period changes, when increasing or decreasing, look like secular changes, and it had been feared that if these perturbations were intrinsically secular, they would eventually cause Saturn to drift away from the solar system and Jupiter to fall into the Sun. Extending his approach, Laplace was able to demonstrate the stability of the solar system by proving that all the relevant planetary perturbations were periodic and that the dimensions of the planetary orbits would stay within certain bounds.‡

If this result seemed a reassuring consequence of celestial mechanics, then the discovery of Neptune in 1846 was a more dramatic proof of the powers of this area of mathematics. Quite independently of each other, but both responding to well-known troubles with the orbit of Uranus, John Couch Adams in Cambridge and Le Verrier in Paris predicted the existence of a previously unknown planet outside the orbit of Uranus. Both had to cope with the inverse problem of perturbation theory, that is finding an unknown body from known perturbation effects rather than finding the perturbation effects of a known body. By a combination of circumstances that enraged English feelings, it was Le Verrier's prediction that caused a Berlin astronomer to discover the planet. Subsequently national feelings were overcome and honour was given to both men, with the new planet being given the neutral classical name of Neptune. The American Benjamin Peirce spoiled things slightly by suggesting that the discovery was due to chance, but generally it was hailed as a triumph of celestial mechanics.

By this time another area of theoretical astronomy, lunar theory, had

† This will be discussed in greater detail in 5.2, Chapter 5.

‡ The nature of Laplace's proof was questioned towards the end of the nineteenth century. The stability of the solar system is now only held to have limited meaning.

largely been conquered. Lunar theory was much more difficult than planetary theory, and had evolved techniques of its own, because the main perturbation of the Moon is due to the Sun, which produces a much larger perturbing force than is found in the planetary theories. Mention has been made of the troubles with the lunar perigee in the first half of the eighteenth century, which almost caused Clairaut and others to abandon the inverse square law, but which were cleared up in the years surrounding 1750. By 1765 the lunar theory was so good that a posthumous award was made by the British Parliament to Tobias Mayer, a German astronomer who had with the theoretical aid of his compatriot Euler drawn up tables of the Moon's motion sufficiently accurate to enable terrestrial longitudes to be found to within 1° when at sea.

There was, however, a continuing problem in the lunar theory—the secular acceleration of the Moon's mean motion, discovered by Halley in 1695. Euler ascribed the acceleration to a resisting ether in space, but Laplace in 1787 was able to show that it could be accurately accounted for by Newtonian perturbation theory. It was another triumph for the theory and another for Laplace. Laplace also worked on another problem, discovered by Burg in 1798, arising from discrepancies in the Moon's mean longitude. He suggested half-heartedly the existence of two long period inequalities, one due to the Sun and the other to an asymmetry of the Earth about the equator. In 1835 Denis Poisson confirmed the suspicions that these two inequalities were insensible and therefore not the cause of the anomaly, which was left unexplained. Trouble was averted by P.A. Hansen, who found two totally unexpected inequalities due to the action of Venus. Having related this saga in his classic *History of Physical Astronomy* of 1852, the historian and astronomer Robert Grant found himself able to add:

> The lunar theory may, therefore, now be considered as divested of all serious embarrassment; and in its present state it undoubtedly constitutes one of the noblest monuments of intellectual research which the annals of science offer to our contemplation. (Grant (1852) p. 121)

Alas for Grant, within a year the lunary theory was in trouble again. In 1853 John Couch Adams showed that Laplace had not taken a sufficient number of terms in the series solutions in his work on the Moon's secular acceleration, and by 1859 he had shown that only half the observed amount was due to perturbations. The remaining half was soon put down to tidal friction, but this was not placed firmly on a quantitative footing until the twentieth century. Agnes Clarke, in the introduction to her *History of Astronomy in the Nineteenth Century* of 1885 could still write:

> The advance of astronomy in the eighteenth century ran in general on an even and logical course. The age succeeding Newton's had for its special task to demonstrate the universal validity, and trace the complex results of the law of gravitation. The accomplishment of that task occupied just one hundred years. It was virtually brought to a close when Laplace explained to the French Academy, November 19, 1787, the

cause of the moon's accelerated motion. As a mere machine, the solar system, so far as it was then known, was found to be complete and intelligible in all its parts; and in the Mecanique Celeste its mechanical perfections were displayed under a form of majestic unity which fitly commemorated the successive triumphs of analytical genius over problems amongst the most arduous ever dealt with by the mind of man. (Clerke (1885) pp. 2–3)

That Agnes Clerke was able to write this thirty years after Adams had shown that Laplace's work was deficient, illustrates the contemporary feelings toward celestial mechanics. When Laplace explained the Moon's motion in 1787, it was taken as a great triumph, both for him and Newtonian theory, but when Adams's papers were published and accepted, they were praised but were not taken as a great refutation of Newtonian theory. This was partly because Newton's law was by the mid-1850s so well confirmed that one anomaly in it would not have seriously dented people's belief in it. Also such discrepancies could be the signpost to new and important discoveries, as for example troubles in the orbit of Uranus pointed to the existence of Neptune. It was hoped that the lunar acceleration would lead to something new, though in this case there was the notion of tidal friction ready to hand.

To Clerke, Mercury's perihelion advance was a 'pending problem' (Clerke (1885), p. 299) because the hoped for discovery of very small planets between Mercury and the Sun had by then come to nought, and the other possible solutions were in stalemate. Eventually it was the physicists rather than the astronomers who gave the final explanation of the anomalous advance in the perihelion of Mercury. The physicists concerned were radicals with no firmly established commitment to Newton's inverse square law and who had questioned such associated tenets as absolute space. The astronomers meanwhile had adopted their own stopgap solution to the problem, just as they had embraced tidal friction to explain what perturbations could not. Once they had accepted the physicists' viewpoint, the stopgap solution was swiftly dropped and planetary perturbations, albeit with relativistic additions, gave the complete perihelion advance for Mercury.

2

LE VERRIER AND THE ANOMALOUS ADVANCE IN THE PERIHELION OF MERCURY

2.1. Le Verrier and Mercury

Urbain Jean Joseph Le Verrier was a scientist of high ambition. Born in 1811, he was educated at the prestigious École Polytechnique in Paris. He started work as a chemist in the civil service and then returned in 1837 to the École Polytechnique as an assistant in astronomy. Le Verrier changed from chemistry to astronomy in order to obtain the valued post as an assistant (see Grosser 1962); his marked ability at mathematics made the move painless and he soon won attention with work on the stability of the solar system. In 1840 François Arago, the director of the Paris Observatory and as such the French equivalent of the English Astronomer Royal, suggested to Le Verrier that he work on the subject of Mercury's motion. The young scientist, eager to oblige, immediately began the lengthy process of constructing the theory. He published a provisional theory in 1843 which was tested at the 1848 transit of Mercury. Unfortunately there was not close agreement between the observations and Le Verrier's predictions, and he reflected on the possible reasons (Le Verrier (1849) p. 2):

> If the tables [of Mercury's positions] do not strictly agree with the group of observations, we will certainly not be tempted into charging the law of universal gravitation with inadequacy. These days, this principle has acquired such a degree of certainty that we would not allow it to be altered; if it meets an event which cannot be explained completely, it is not the principle itself which takes the blame but rather some inaccuracy in the working or some material cause whose existence has escaped us. Unfortunately, the consequences of the principle of gravitation have not been deduced in many particulars with a sufficient rigour: we will not be able to decide, when faced with a disagreement between observation and theory, whether this results completely from analytical errors or whether it is due in part to the imperfection of our knowledge of celestial physics.

Le Verrier held back from blaming the inverse square law of gravitation or some unknown 'material cause'—i.e. unknown planetary matter—on the grounds that this 1843 theory was not sufficiently exact. By 1849 the discovery of Neptune had given the Newtonian gravitational law a big boost and Le Verrier instant fame and honours. In 1854 his scientific standing was such that he was chosen to succeed Arago as director of the Paris Observatory.

2.2 The anomalous perihelion advance

In 1859 Le Verrier produced a more sophisticated theory. His Théorie du mouvement de Mercure (Le Verrier 1859*a*,*b*) was the most detailed work possible, and was sufficiently rigorous for any disagreement with observation to be taken quite confidently as indicating a new scientific fact.

Theories such as the one Le Verrier produced for Mercury consisted of two parts, observational and theoretical. A set of observations was used to test the theoretical expressions; these described the elements of the planet's orbit and contained terms due to perturbations by other planets. The aim was to produce agreement between the two parts. The differences between the theoretical and observational terms were known as the residuals, and the elements were varied in order to minimize the residuals. Differences between observation and theory had to be less than observational errors, which by the middle of the nineteenth century were of the order of 1″ in measuring the position of a planet. If larger differences remained, some of the theoretical variables had to be changed. Typically a planetary mass had to be altered, but whatever variable was changed it had to be done in such a way that those parts of the theory that had been satisfactory remained like that. If this could not be achieved, then the offending variable or element had to be identified and the error explained.

Le Verrier used two sets of observations in his 1859 theory. The first comprised a series of 397 meridian observations of Mercury taken at the Paris Observatory between 1801 and 1842. Comparison of these with the provisional tables from the theory showed errors of a few seconds of arc. The second set of observations were of 14 transits of Mercury, which though many fewer in number were by their nature more accurate. The observations were of internal contacts on entry and exit of the planet as it crossed the Sun's disc. As we have seen in Section 1.3, transits of Mercury can take place in either May or November. Le Verrier found that the residuals between theory and observation for the November transits were of the order of 1″ and had no regular pattern. The residuals derived from the May transits were quite different, and had the values given in Table 2.1. The values prompted Le Verrier to comment (Le Verrier (1859*a*) p. 78):

> Ignoring the observations of 1661 and 1677, we notice that the observations of transits at the ascending node (November) only give rise to small errors; whereas the transits at the descending node (Mercury) give rise to an error of 12″.05 in 1753 which is reduced to −1″.03 in 1845, having decreased almost steadily with time.
>
> These 13 seconds of variation in 92 years must be taken seriously due to the accuracy of the method by which they were observed. In fact they cannot be ascribed to uncertainties in the observations of the transits, as we would have to imagine that all the astronomers had committed errors increasing with time and differing by several minutes over the 92 year period. We cannot allow that!

Le Verrier had found discrepancies between theory and observation that were

TABLE 2.1

Year	Entry	Exit
1661.33	$+12''.7$	—
1753.34	—	$+12''.05$
1786.34	$+4''.84$	$+5''.11$
1799.34	$+5''.65$	$+3''.83$
1832.34	$+0''.17$	$-0''.58$
1845.35	$-1''.03$	—

larger than could be explained by observational error alone, and had to return to the theory to find the cause. He obtained an equation related to one that was derived in Section 1.3, namely eqn (1.8):

$$L = l + 2e \sin(l - \varpi).$$

Differentiating this with respect to time, and substituting standard values for l, e and ϖ, gives an expression (see eqn (1.11)) relating $\mathrm{d}e/\mathrm{d}t$, $\mathrm{d}\varpi/\mathrm{d}t$ and $\mathrm{d}l/\mathrm{d}t$. The value of the latter term is derived from observation. Le Verrier proceeded along similar lines. To a higher order eqn (1.8) is:

$$L = l + 2e \sin(l - \varpi) + 5/4\, e^2 \sin(2l - 2\varpi) + \ldots \tag{2.1}$$

Using the method of variation of elements and taking all the transit observations as his data Le Verrier found the following equation (see Tisserand (1896) pp. 520f.):

$$2.72\, \mathrm{d}e/\mathrm{d}t + \mathrm{d}\varpi/\mathrm{d}t = +0''.392. \tag{2.2}$$

In this equation $\mathrm{d}e/\mathrm{d}t$ and $\mathrm{d}\varpi/\mathrm{d}t$ are the changes measured in seconds of arc per year in the eccentricity and perihelion of Mercury which were required in addition to the known perturbations to account for the transit observations. From one equation two variables cannot both be determined, and to do this Le Verrier turned to the other set of observations. He used the Paris meridian observations of Mercury in two groups, corresponding to the years in which they were carried out, of 1801–28 and 1836–42. These gave $\mathrm{d}e/\mathrm{d}t = -0''.0743$ and $\mathrm{d}e/\mathrm{d}t = -0''.0869$ respectively. Substitution of the mean $\mathrm{d}e/\mathrm{d}t$ in eqn (2.2) gave $\mathrm{d}\varpi/\mathrm{d}t = 0''.60$ which was a 'considerable annual variation' (Le Verrier (1859*a*) p. 95). Le Verrier based this remark on the changes in perihelion and eccentricity due to the perturbations of the other planets, which amounted to $5''.267$ and $0''.042$ per year respectively as we have seen in Section 1.3. A correction to the perihelion of $0''.60$ per year was a 10 per cent increase, a value too large to accept, and Le Verrier abandoned his attempt to try to determine the eccentricity change separately.

An explanation now had to be given for eqn (2.2). Le Verrier decided to find out whether the quantities could be obtained by increasing the mass of Venus, the nearest neighbour of Mercury and which already produced most of its advance in perihelion. He found that a 10 per cent increase in the mass of Venus produced the perturbations required to satisfy eqn (2.2). However this extra mass also produced additional perturbations in the other neighbour of Venus, namely the Earth. These proved quite unacceptable, especially in a quantity known as the obliquity of the ecliptic which had been well determined.

Le Verrier now knew it was impossible to use known matter to produce the anomalous perihelion advance, and he had to try the possibility of new undiscovered matter. This was what he had had in mind in his 1849 remark about a material cause, and was a repetition of the strategy that had led to the discovery of Neptune in 1846. Since the new matter had to perturb Mercury but not Venus, as this planet's theory was relatively satisfactory at this stage, it had to lie between the orbit of Mercury and the Sun. There were two possible forms for this intra-Mercurial matter—a single planet or a number of small minor planets such as had recently been discovered between Mars and Jupiter. In order to determine more about the hypothetical matter, Le Verrier gave the following expressions for the variation in the perihelion ϖ and eccentricity e of Mercury due to a perturbing mass m':

$$\mathrm{d}\varpi/\mathrm{d}t = \frac{1}{2}\left\{B + C\frac{e'}{e}\cos(\varpi - \varpi')\right\}m'n \tag{2.3}$$

$$\mathrm{d}e/\mathrm{d}t = \tfrac{1}{2}\{C\,e'\sin(\varpi - \varpi')\}m'n. \tag{2.4}$$

In these equations B and C are coefficients depending on the ratio of the semi-major axis of the orbits of the perturbing matter and Mercury, n is the mean motion of Mercury, and e' and ϖ' are the eccentricity and perihelion of the new matter.

Le Verrier also obtained a modified form of eqn (2.2):

$$\mathrm{d}\varpi/\mathrm{d}t + 2.72\,\mathrm{d}e/\mathrm{d}t = 0''.383. \tag{2.5}$$

Now Le Verrier assumed, in order to make the problem tractable, that the orbit of the hypothetical matter had a negligible eccentricity, and put $e' = 0$. This was certainly an improbable value, for the next nearest planet to the Sun, i.e. Mercury itself, had a large eccentricity of about 0.2, and the eccentricities of the minor planets known then were mostly between 0.1 and 0.2. However this step dramatically simplified the problem, for from eqn (2.4) Le Verrier found that $\mathrm{d}e/\mathrm{d}t = 0$. Substituting this into eqn (2.5) gave $\mathrm{d}\varpi/\mathrm{d}t = 0''383$ per year or 38″.3 per century.

At this point the 38″ per century anomalous advance in the perihelion of Mercury entered the history of science. Le Verrier next had to determine the quantity of matter that gave this 38″ advance, and found that having put $e' = 0$ his task was straight forward. Equation (2.3) now reduced to:

$$d\varpi/dt = \tfrac{1}{2} Bm'n$$

or

$$Bm'n = 0''.766 \text{ per year.} \tag{2.6}$$

Le Verrier then used the function B to give a set of possible masses and distances for the new planetary matter (Le Verrier (1859) p. 104): there are set out in Table 2.2.

TABLE 2.2

Distance of planet to Sun (au)	Ratio of planet's mass to that of Mercury	Greatest angular distance from Sun
0.310	0.07	18° 4′
0.271	0.17	15° 43′
0.232	0.35	13° 25′
0.194	0.68	11° 11′
0.155	1.29	8° 55′
0.116	2.66	6° 40′

As a comparison, Mercury is 0.39 au from the Sun with a greatest angular distance of 28°.

Le Verrier continued by discussing the observational consequences of his hypothesis of intra-Mercurial matter. A single planet of the required size should have been observed already. The nearer to the Sun the planet was, the larger it had to be to produce the full perturbing effect on Mercury; the camouflage of orbiting near the brilliant light of the Sun itself was balanced by the larger and more visible size required. Such a planet had never been observed during total eclipses of the Sun, nor had a transit been observed despite frequent sunspot searches. The absence of any reported sightings led Le Verrier to consider the alternative (Le Verrier (1859*a*) p. 105):

Those to whom these objections appear too serious will be led to replace this single planet by a series of asteroids whose effects add up to produce the same total result on the perihelion of Mercury. Apart from the fact that these asteroids will not be visible under ordinary circumstances, their distribution around the sun will be such that they will introduce no important periodic inequality in the motion of Mercury.

The hypothesis to which we have been led is in no way extreme. A group of asteroids is found between Jupiter and Mars, and to be sure we can only make out the main ones individually. There is even reason to believe that planetary space contains an unlimited number of very small bodies in motion round the sun: for the region neighbouring the earth's orbit that is certain. The continuation of observations of Mercury will show whether we can definitely admit that such groups of asteroids also exist much nearer the sun.

On this point Le Verrier closed the discussion in the *Théorie du mouvement de Mercure*. His results and hypotheses were immediately published in the French *Comptes Rendus de l'Académie des Sciences* in the form of a letter to Faye, the Secretary of the *Académie des Sciences* (Le Verrier 1859*b*). Soon Le Verrier received a letter announcing something that all his writings indicated that he thought would not happen: an astronomer had observed the transit of an intra-Mercurial planet. A country doctor called Lescarbault, an amateur astronomer, had observed the transit of a planetary object earlier in the year. Le Verrier travelled to Dr Lescarbault's home in Orgères, south-west of Paris, made inquiries, became convinced of the doctor's good faith and of the adequacy of his apparatus, and accepted the observation as that of an intra-Mercurial planet which was named Vulcan (Le Verrier (1859*a*) pp. 394–9). In an account of the discovery in the *Monthly Notices of the Royal Astronomical Society*, it was compared to the discovery of Neptune which had occurred as a result of Le Verrier's predictions in 1846 (Anon. (1859–60) p. 100):

> The singular merit of M. Lescarbault's observation will be recognised by all who examine the attendant circumstances; and astronomers of all countries will unite in applauding this second triumphant conclusion to the theoretical inquiries of M. Leverrier.

But not every astronomer applauded Le Verrier. In 1860 the French astronomer Liais, who was working with the Brazilian Coast Survey, reported that he had been observing the Sun at the same time as Lescarbault's alleged sightings of Vulcan and had seen no sign of it. Liais had been measuring the decrease in luminosity of the solar disc from the centre to the edge and from the solar equator to the poles. He could not believe that he could have missed seeing Vulcan had it crossed the face of the Sun (Liais (1860) col. 370):

> Now is it possible that in carrying researches on the physical constitution of the sun with a magnification double that of M. Lescarbault I would not have observed a sunspot at 79° from the equator, when for each comparison I examined carefully the solar area to take account of apparent variations in intensity of the wrinkles and ridges.
>
> Consequently I am in a position to deny in the most clear and positive way the passage of a planet across the sun at the time indicated.

Liais was adamant that Lescarbault was mistaken. He did not hold much respect for Le Verrier's suggestions either. He pointed out that a planet between Mercury and the Sun would have a visible disc, and that none had been observed during a solar eclipse or under ordinary conditions. He added that the 38″ advance in Mercury's perihelion could be decreased if account was taken of refraction affecting observation of the planet, and that a smaller value could be explained by increasing the mass of Venus. Liais thought that the effect of this on the obliquity of the ecliptic did not matter as astronomers generally overestimated the precision of their observations. In a later treatise on astronomy, Liais (1881, p. 476) recalled this episode and the paper he had written:

> This memoir brought me congratulations from several illustrious scientists. One of the most celebrated astronomers in Europe even wrote to me a flattering letter in which he added that I was well thought of by scientists in opposing the light-headed fantasies concerning the existence of a planet between Mercury and the sun.

Unfortunately the astronomer was not named. Three pages earlier Liais (1881, p. 473) had revealed his antipathy to Le Verrier; in describing the events leading to the discovery of the planet Neptune in 1846 which had been made by Galle using predicted positions provided by Le Verrier, he concluded:

> To Galle therefore, and not to Le Verrier, the honour of the discovery, as to Newton, and not to the apple, that of universal gravitation.

Liais had drawn the lines for a scientific battle. As we shall see, the battle never developed into a good debate and was to a certain extent irrelevant. But first we must digress slightly, for Le Verrier was not alone in suggesting the existence of intra-Mercurial matter. Three other nineteenth century scientists did the same, all for completely different reasons. Their stories predate but also mingle with that of Le Verrier's Vulcan, and it is evident that the idea of intra-Mercurial matter had been widely aired before Le Verrier discovered the anomalous advance in the perihelion of Mercury.

2.3. Babinet's Vulcan

The name Vulcan had been put forward as the name of an intra-Mercurial planet by the French astronomer Babinet (1846). He had a penchant for this type of activity, for in 1848, when it was apparent that the elements of the newly discovered Neptune were not in total agreement with those of the predicted planets of Le Verrier and Adams, Babinet suggested that there was a planet further from the Sun than Neptune which he named Hypérion (Babinet 1848). He adjusted the elements of Hypérion to account for the difference between those of Neptune and the predicted planet; thus its mass was the difference between that of Neptune and the mass predicted by Le Verrier. This showed a lack of subtlety but there was at least observational evidence for Babinet's Vulcan. During the solar eclipse of 1842 prominences had been observed around the Sun's disc, and though these had been seen before their status as probable optical illusions had pre-empted the requirement for physical explanation. The 1842 eclipse had provided a problem, however, for Mauvais and Petit had observed an incredible growth of a prominence which had increased in height from 1′17″ to 1′45″ in angular measure during an eclipse which had a duration of only 2 minutes. Since the Sun's diameter is about 32′, this represents an increase in altitude of about 10^4 km which is a large amount in the strong gravitational field of the Sun. The problem was to find what could possibly be the force behind this outburst if indeed it was caused by an ejection of solar material.

Babinet's suggestion was that it was partly due to an illusion, and that the appearance of prominences was due to 'incandescent clouds of a planetary kind, circling the Sun in the form of a train or portions of a ring' (Babinet (1846) p. 282). Being of the same colour as the surface of the Sun, they would be hard to observe while passing across its face but would be seen at the sides as an ascending and descending flame, and, if the ring portion was long enough, the flame would never appear detached. A period of revolution of 4 hours would be sufficient to explain the observations of Mauvais and Petit. This was an improbably short period since the period of rotation of the Sun on its own axis is about 608 hours and no bodies were known which had periods shorter than the period of revolution of their primary until Asaph Hall's discovery of the satellites of Mars in 1877. Babinet cited the comet of 1843 (the Great comet of 1843) as passing behind the Sun, from node to node, in just over 2 hours. This at least showed that such high speeds were possible, though the analogy could not be taken too strongly for the period of this Great comet was about 175 years! The clouds would originate either from conglomerations of cometary matter or as the final remnants of the cooling and condensing gas from which the planets had formed on the lines of Laplace's nebular hypothesis. For the naming of the matter (Babinet (1846) p. 282):

> From the nature and appearance of this planetary mass we have to give it the mythological name of Vulcan and call other analogous masses that might be specified the Cyclopes.

Babinet added that this theory found little favour with those to whom he showed it, though we are not told what the reasons were or who the people were. Meadows (1970) writes concerning prominences, that 'they were, by general agreement, considered to be clouds floating above the solar surface'. Faye continued to believe that prominences were the result of optical illusion. It was not until 1868 that Lockyer and Janssen developed spectroscopic methods of observing prominences at times other than solar eclipses, and up to this time no hypothesis could be more than vague. Even then prominences, and especially the enormous forces displayed when one burst forth, were difficult to explain. One might have objected to Babinet's clouds on the grounds that with the high speeds involved they would fall into the Sun under the influence of the resisting medium put forward to explain the acceleration of Encke's comet. Also they have insufficient mass, being gaseous, to perturb Mercury's orbit sensibly. Nevertheless we have in Babinet's Vulcan and Cyclopes shadowy ancestors of Le Verrier's Vulcan and asteroid ring.

2.4. Buys-Ballot's rings

The next example of an intra-Mercurial planet is more substantial but with a less than firm empirical basis, though one that was quite strong enough for its author and indeed for several other later workers. The Dutch meteorologist C.H.D. Buys-Ballot is today remembered only for the Buys-Ballot law which states that, for the northern hemisphere, if one stands with one's back to the wind the region of lowest pressure is on one's left. In the nineteenth century Buys-Ballot was the leading meteorologist in Holland and was responsible for the initiation of large-scale weather observations and interpretations. One of the preoccuations of Buys-Ballot was the existence of periodic variations in the temperature of the atmosphere. Such variations were quite often discussed at this time, though the only one to achieve wide acceptance was the 11-year periodicity in the sunspot numbers found by Schwabe in 1843. Buys-Ballot's period for the temperature variation was short, being $27.684 \pm 0{\cdot}005$ days (Buys-Ballot 1846, p. 206). The parameter which suggested itself as responsible for the apparent variation was the rotation period of the Sun. If, for instance, the Sun's surface were not uniformly hot and just one area were the hottest, one would expect a temperature variation with the same period as that of the Sun's rotation which showed a maximum when, or just after, the hottest part was facing the Earth. The Sun's rotation period was measured by using sunspots as points fixed to the Sun's surface, but the value obtained by this method was 27.23 days, slightly smaller than Buys-Ballot's period, and he could not make the smaller period fit his temperature data. One could of course have refused to take the sunspot period as a perfect measure of the rotation period of the Sun itself. Though Buys-Ballot tried to adopt this position, he realized that one would have expected, if anything, a sunspot period that was longer than the rotation period. A shorter period would mean that the sunspots were travelling faster than the surface of the Sun, whereas on grounds of frictional drag one might expect the opposite, unless one held that the sunspots were fixed to the surface of the Sun in which case the two times would be equal. Buys-Ballot was therefore forced to adopt the last position and to find a new explanation for his temperature period. Citing as a guide *sine hypothesi scientia nulla* with a rider 'Hypotheses only hurt when one forgets that they are hypotheses', he proposed that there was a ring of matter surrounding the Sun. This had a rotation period of 27.68 days and, being less dense in certain parts so that there was differential scattering of the solar radiation, produced a 27.68 day period in the Earth's atmospheric temperature. Buys-Ballot cited as supporting analogies the rings of Saturn and the zodiacal light, which was generally held since the time of Cassini to be light reflected off many small particles circling the Sun.

In 1847 Buys-Ballot published a more extensive paper based on temperature measurements taken between 1729 and 1846 (Buys-Ballot 1847). In monograph form, this seems to be a rare work and in a paper published in 1860 the section discussing the ring hypothesis was reprinted (Buys-Ballot 1860). He introduced a second ring of period 27.56 days to account for another periodic variation in temperature that he had found, and the two rings formed analogous counterparts to the two rings of Saturn as well as being possible explanations of the solar prominences.

The prediction of intra-Mercurial matter by Le Verrier in 1859 led Buys-Ballot to write to him about his own hypothesis. One cannot say whether Le Verrier knew of Buys-Ballot's rings, though he must have known of his work as a fellow meteorologist. We have at least the astronomer's reaction, recorded in the journal *Cosmos* (1859) from a meeting of the French Académie des Sciences:

> M. Leverrier announced formally that in no way did he share his opinion; that his rings did not satisfy theoretical conditions; that he could not even admit the actual existence of parts of a disjoint ring, of segments becoming disjoint in time or being dissipated in space or blending into a single ring etc. It was none the less true that M. Buys-Ballot had admitted long ago the existence of planetary rings inside the orbit of Mercury.

Le Verrier's reply, coming as it did from celestial mechanics, probably would not have occurred to Buys-Ballot and it does not seem to have convinced him since he continued to publish papers on his rings. The instability of a disjointed ring was not the only objection given. Airy (1855) and d'Arrest (1854) for example, both declared that the temperature data just did not support such periodic variations as Buys-Ballot wished to show. Of course any set of data was able to be interpreted using periodic terms on the lines of Fourier series and some sets of periodic terms were going to give a better fit than others, but it was not then said that these were the real variations underlying the data. However, a single periodic variation was more likely to be the true variation if it gave a good fit, and on these grounds Schwabe's 11-year sunspot period was accepted. Buys-Ballot spoiled his case with the introduction of the second period and the resulting improvement to his fit.

The sufficiency of the proposed mechanism of the rings was also questionable. It was doubtful whether a ring of matter which could not be observed and which was not everywhere equally dense could produce a temperature variation sensible in the Earth's atmosphere. For a homogeneous ring, the rings of Saturn could have been used as an analogy. Since they were inclined to the ecliptic they varied in their apparent brightness and were practically invisible every 15 years, the period of Saturn being about 30 years. Now if the rings were not rotating with the Sun they would be invisible only every six months, which was too long for Buys-Ballot's purpose. If the ring were attached to the Sun it would have been invisible every 13 or so days, which was too short. The solution, a ring inclined to the ecliptic but which rotated around an axis which was perpendicular to the ecliptic with a period of 55.364 days (which was inside the orbit of Mercury), was dynamically unacceptable.

Later work in this field avoided the difficulties of the mechanism by bringing in sunspots. The 11-year sunspot cycle was already established as the cause of variations in temperature, variations in pressure, famine, etc., and it was conceivable that there might exist short-period variations in sunspot numbers quite apart from the rotation period. If sunspots were regions of intense solar activity, short-period variations in their numbers would be reflected in short-period variations in the effects of solar activity, such as temperature. A possible mechanism for the variation in numbers was considered to be the tidal influence of the planets as they passed round the Sun. These tides obviously existed, but to an uncertain extent, and it was suggested that a large tidal wave, produced say by the conjunction of two near planets, might act as a trigger and produce a sunspot. Sunspot numbers then would exhibit periodic variations which could be correlated with the planetary movements. The large 11-year cycle could possibly be explained by the 11.86 year period of the largest planet Jupiter, with better agreement being obtained by introducing the extra influence of Saturn. Workers in this field included the observers at the solar observatory at Kew, Warren de la Rue, Balfour Stewart, and B. Loewy, and in 1881 Stewart read a paper at the annual meeting of the British Association entitled *On the possibility of the existence of intra-Mercurial planets* (Stewart 1881). They had found maxima/minima in sunspot numbers corresponding to the perihelion/aphelion of Mercury and conjunction/opposition of Mercury and Jupiter, Venus and Jupiter, and Venus and Mercury. Stewart announced the detection of a period of 24.011 days and the possibility that this might be due to an intra-Mercurial planet. To test this he had sought periods corresponding to the synodic periods of this planet with Mercury, Venus, and Jupiter, i.e. periods of 33.025, 26.884, and 24.145 days respectively. Good agreement was obtained as Steward found prominent periods of 32.955 and 26.871 days and a less prominent period of 24.142 days. However, he then went further and by equating the maxima of these periods with conjunctions he found three independent values for the longitude of the intra-Mercurial planet at a particular epoch. These did not agree very well, the largest discrepancy of one of the three values being 20° from the mean, but Stewart was optimistic:

> It would appear from this investigation that the evidence is in favour of the sun-spot inequality of 24.011 days being due to an intra-Mercurial planet. (Stewart 1881, p. 464)

There was no mention of the consequences of this for the motion of Mercury. At this time there were no major supporters of the single planet hypothesis for, as we shall see below, this was observationally untenable on optical grounds by the late 1870s and any proponent of the hypothesis using as supporting evidence observations other than optical, such as Stewart was doing, had to contend with this negative evidence. Explanation involved recourse to improbably high densities, the values of which were approximately equal for the four inner planets. However, Stewart did not have to go this far for his own supporting evidence was very weak. We have already criticized the search for periodic variations in sets of data, especially when many periods were found. Stewart's periods already numbered eight. This particular investigation does not seem to have been followed up, but the type of work continued. A late intra-Mercurial planet was invoked in Huntington's *Earth and sun* (1923) which discussed planetary influences in connection with periodic variations in sunspot numbers and weather phenomena. Huntington suggested that this planet might explain the advance in the perihelion of Mercury, which might seem rather a late if not redundant suggestion to make but was one that fitted into a certain school of thought of the time. We shall mention later Charles Lane Poor, whose *Relativity versus gravitation* was published in 1922. In this book Poor tried to show how suitable arrangements of intra-Mercurial matter could make Newtonian gravitational theory far more successful than general relativity in explaining both the phenomena of the solar system and the deflection of starlight in the gravitational field of the Sun.

2.5. Kirkwood's analogy

Our final example of intra-Mercurial planets being suggested before 1859 is concerned with the work of Daniel Kirkwood. Kirkwood rose to sudden fame in 1849 when he announced the empirical law known as Kirkwood's analogy which stated that the quantity n^2/D^3 was a constant for all the planets, where n was the number of rotations performed by a planet in each revolution round the Sun and D was the distance between the two points on either side of a planet at which the attraction was divided equally between that planet and its respective neighbour, all three planets being in conjunction (Kirkwood 1850). The 'analogy' had connections with the nebular hypothesis of the origin of the solar system proposed by Laplace. One might have expected that during the condensation and formation of the planets some matter would have remained left over—possible asteroids, meteors, or as yet unobserved debris in the solar system. Intra-Mercurial matter would have been quite plausible on this account, and Babinet did not hesitate to cite it as a possible origin of his Vulcan and Cyclopes. However, the analogy itself was based on slender evidence. Of the eight planets, the quantity D could not be determined for Mercury or Neptune which had only one neighbour, nor for Mars or Jupiter which had the asteroids between them, nor for Uranus whose period of rotation was not properly known. This left Venus, the Earth, and Saturn, whose elements approximately fitted the analogy, to support it. The paucity of the data, which was pointed out by B.A. Gould (1850) among others, was not apparent to Sears C. Walker who was well known for having been the first to obtain the elements of Neptune from the early observations of that planet. Walker was the person who gave Kirkwood enough confidence to publish his analogy, and in order to improve the appearance of the analogy introduced a planet called Kirkwood into the gap between Mars and Jupiter. Kirkwood (the planet) fitted the earlier hypothesis of Olbers that there had originally been a planet at a mean distance of 28 au (to fit Bode's law) which had exploded and that the fragments had become the asteroids. The mean distance of the asteroid planet according to Kirkwood's analogy was 29 au, in fair agreement with Bode's law. Though the evidence for the analogy was weak, it was taken very seriously, especially in America where it was compared to Kepler's third law, and it was hoped that someone would link it firmly to the nebular hypothesis and so play Newton to Kirkwood's Kepler (see Numbers 1973).

Both Kirkwood and Walker saw the possibility of an intra-Mercurial planet, and Walker gave it the elements that fitted the analogy—a mean distance of 0.20 au and a mass 1/4 739 670 where the Sun's mass is 1 but

TABLE 2.3

Pair	Planet	Mean Diameter	Density
I	Neptune	4.739	0.187
	Uranus	4.428	0.153
II	Saturn	9.205	0.133
	Jupiter	11.255	0.243
III	Asteroid planet	0.584(?)	1.472(?)
	Mars	0.519	1.032
IV	Earth	1.000	1.000
	Venus	0.991	0.973
V	Mercury	0.391	1.930
	—	—	—

From Kirkwood 1852, p. 217.

asserted no more than the possibility of its existence (Walker 1850, p. 22). These elements agreed fairly well with Le Verrier's later requirements. In 1852 Kirkwood offered further clues to the existence of an intra-Mercurial planet (Kirkwood 1852). Using the asteroid planet he showed that the planets could be arranged in pairs using as parameters their density and mean diameter. The pairing is shown in Table 2.3. The arrangement in terms of diameters was quite striking, although the asteroid planet was hypothetical, but all the densities showed was that the four inner planets were very different from the four outer planets. The pairing did suggest a companion to Mercury; indeed if it did not exist this had to be explained, since the connection with the nebular hypothesis was that from each primitive ring two planets were formed either side of the mean circumference of the ring. Kirkwood's paper was favourably reviewed in *Cosmos* (1852), the author writing:

We freely acknowledge that Kirkwood's analogy fascinates us a great deal, and we easily accept either the intra-Mercurial planet or its equivalent in a condensed ring of meteors or shooting stars.

In 1864 Kirkwood was less emphatic on the new planet and introduced a reason why there might not be a companion to Mercury (Kirkwood 1864). This was based, not on his analogy, but on a Bode-type law applied to the radii of gyration† of the primitive rings from which each pair of planets had been formed. The differences between these radii formed a geometric series which gave 3.0980 for the distance of the asteroid planet, a value which was confirmed by the value of 3.1116 obtained using the analogy. Now on these grounds, the radius of gyration of the Mercurial primitive ring was 0.403 35, whereas the radius of the orbit of Mercury itself was 0.419 61:

The radius of gyration of the fifth primitive ring corresponds very closely with the mean distance of Mercury. This planet, therefore, according to the hypothesis, ought to be an exception to the binary arrangement. Such, in fact, appears to be the case. Or, if the planet originally existed as a binary ring, the mean distances of the members having the same ratio to each other as those of Venus and the Earth, both must have been included between the present limits of Mercury's orbit. The union of the two rings and the formation of a single planet may thus have resulted from the eccentricity of the primitive annuli. (Kirkwood 1864, p. 13)

It says much for the poverty of the Vulcan observations that Kirkwood with his belief in the possibility of an intra-Mercurial planet and after Le Verrier's prediction of such a planet, should say that it 'appears to be the case' that there was no such planet. However, the possibility was still left open that there was matter in non-planetary form:

. . . may this slight and only exception to the strict accuracy of the law be referable to zones or groups of asteroids in the vicinity of Mercury's orbit, the existence of which has been indicated by the researches of Leverrier? (Kirkwood 1864, p. 13)

This support for Le Verrier's asteroid ring hypothesis was not all that Kirkwood wished to provide for he obtained further terms from his series. A sixth ring had a radius of 0.262 89, the limit of the series being 0.202 99. Between these two distances the

formula indicates the abandonment at the solar equator of an indefinite number of rings in close proximity to each other. The appearance of such zones or rings of nebular matter would be similar to that of the zodiacal light. This phenomenon was ascribed by Cassini to the blended light of an innumerable multitude of extremely minute asteroids revolving round the centre of our system. (Kirkwood 1864, p. 14)

Later we shall consider the zodiacal light in its own right as the matter perturbing Mercury. It is remarkable that something as tenuous as this relation of Kirkwood could indicate both an asteroid ring which was also indicated by Le Verrier and the zone of small particles forming the zodiacal

† Kirkwood used this term to designate the mean radius of the ring.

light. However the rings were merely hypothetical, of course, and were based on the pairing arrangement introduced in 1852 which was itself based on the Kirkwood analogy, and this had a slender amount of supporting evidence.

2.6. Reactions to Vulcan

The preceding three sections make it clear that ideas of intra-Mercurial matter had been prevalent and well publicized before Le Verrier made his predictions in 1859. It was therefore not novel that it was intra-Mercurial matter that was being predicted. Le Verrier also steered clear of controversy by not considering the possibility of altering the Newtonian inverse square law. We have seen in Section 1.4 how such an alteration produces a perihelion advance. In 1849 Le Verrier had stated that the truth of the inverse square law had been placed beyond doubt, and in 1859 he did not even discuss it though a quantitative appraisal could have been given without difficulty. This type of solution to the anomalous perihelion advance will be described in Chapters 5 and 6.

Le Verrier had proposed two alternative explanations of Mercury's perihelion anomaly. One was the existence of a single intra-Mercurial planet and the other was the existence of a number of small minor planets or asteroids inside Mercury's orbit. When Lescarbault had disclosed his observation of Vulcan, quick calculations by Le Verrier given in a postscript to his paper Le Verrier (1859*a*) revealed that the mass of the new planet (based on its apparent diameter) could only be about one seventeenth of the mass of Mercury. At the distance from the Sun that also had been calculated for it, Vulcan was far too small to provide the whole of the 38″ advance. Le Verrier therefore required other minor planets, but this need was rather overshadowed. Astronomers were preoccupied with Vulcan. Soon after its discovery Professor Wolf of Zurich found in earlier astronomical literature 21 references to possible planetary bodies transiting the Sun; R.C. Carrington chose eight of these between 1761 and 1820 as being 'deserving of attention' (Anon. (1859–60), p. 101). The observations that indicated planets possibly identical to Vulcan were used in calculating its orbit, and Radau predicted a further transit in the spring of 1860. None of the astronomers who search for Vulcan's transit saw anything. At this time Liais's attack on Lescarbault's observation was published.

Liais was not the only person to object in print in 1860 to the Vulcan hypotheses. The American Simon Newcomb, whose later theory of Mercury's motion superseded Le Verrier's, attacked certain consequences of the idea. He pointed out that since no planets like Vulcan had been observed during solar eclipses, and since even Vulcan itself only gave 1/17 of the required perihelion motion, the intra-Mercurial matter would have to be in the form of hundreds of small asteroids in a ring. The ring would be likely to have a small mean inclination. But a ring of small inclination would affect the nodes of Mercury, as only a ring having the same inclination as Mercury (7°) could leave its nodes unperturbed. Since Le Verrier had found that the nodes of Mercury were not moving anomalously, Newcomb reasoned, the asteroid ring hypothesis could not be true.

Much later, the German J. Bauschinger (1884) published a paper in which he quoted extracts from a letter that Newcomb had sent to him, in which Newcomb retracted this argument. Though the mean inclinations of the asteroid ring over a long time would be zero, at any particular time the inclination might well be similar to that of Mercury. Given a suitable orientation, the ring could then perturb Mercury's perihelion but not its nodes. This 1884 paper and the earlier one of 1860 provoked little comment, and interest in the subject was monopolized by the single planet Vulcan.

Belief in Vulcan refused to die despite the poverty of the observations of it and despite its having a mass that was too small to give much of the 38″ advance in Mercury's perihelion. This belief was sustained by occasional claims of sightings throughout the 1860s and 1870s that have been well documented by Fontenrose (1973). The sightings were largely by amateur astronomers, and none of the organized official searches found anything. Seventeen years after Lescarbault's observation Le Verrier (1876) published a detailed work on Vulcan's orbit. He considered 25 observations of bodies passing across the Sun's disc and accepted 19 of them as reliable. Le Verrier used them to calculate an orbit for Vulcan, and then used that to predict future transits of the planet. But despite diligent searching for almost two decades, few astronomers had claimed to have seen such transits. Le Verrier had to explain both the lack of transits and the seemingly positive sightings, and he was just about able to do this. We have seen in Section 1.2 that transits occur when the transiting planet is at one of its nodes at about the same time as the Earth is at that particular longitude. Le Verrier calculated an orbit for Vulcan that gave a best fit to the observations and showed that transits could take place in March or October. He also showed that for a number of years Vulcan was not at either of its nodes when the Earth was in the required positions, so that transits could not have occurred. But in explaining away the lack of previous transits, Le Verrier committed himself to predicting future ones for the newly calculated orbit. He predicted a 'doubtful' October transit in 1876, but Vulcan was not observed. The next likely autumn transit was to be in 1882, although a March passage was predicted for 1877. Large scale observations found nothing.

On 23 September 1877 Le Verrier died. Within a year a Vulcan controversy started in the United States, when alternate belief and disbelief greeted the separate announcements of James Watson and Lewis Swift that they had both seen a Vulcan-like object during the total eclipse of July 1878. Subsequently it was felt that all the two astronomers had observed were the stars Theta and

Zeta Cancri. One of the opponents of Watson and Swift was another American astronomer, C.H.F. Peters, who in a manner reminiscent of Liais in 1860 criticized the apparent observations of Vulcan (Peters 1879). He admitted that none of his criticisms touched the alternative hypothesis to that of Vulcan—a large number of smaller minor planets—but questioned whether all the 38″ Mercury advance was necessary. Peters suggested that 20″ was sufficient to satisfy the transit observations of Mercury that had led to the discovery of the anomaly, and that of this amount 15″ could be obtained by using a larger value for the mass of Venus. The rest of the discrepancy might then be covered by using a modified law of gravitation analogous to that proposed by Wilhelm Weber for electrodynamics. This law will be discussed in detail in Chapter 6.

Despite the many failed predictions of transits of Vulcan and such arguments as those of Peters, work on intra-Mercurial planets continued. In Berlin Oppolzer (1879*a*,*b*) produced elements of an orbit for Vulcan and predicted a transit for March 1879 which did not occur. In Paris Felix Tisserand had succeeded Le Verrier as the top astronomer in the field of celestial mechanics. In 1882 he published a long survey of the problem of Mercury and the intra-Mercurial planets, ending with these conclusions:

1. It seems to us that we must give up the hypothesis of a single planet producing the established disturbances in the motion of Mercury; this becomes apparent from all the observations made during solar eclipses, and in particular during that of 29 July 1878.
2. If intra-Mercurial planets exist which are comparable in size to the one M. Lescarbault saw pass in front of the Sun, then there cannot be many of them; otherwise they could not have escaped astronomical investigations such as those of Carrington and Spörer who have carefully observed the Sun for about 20 years, describing and measuring the smallest spots which appeared on its surface.
3. These planets could not produce on their own the perturbations in the motion of Mercury; therefore whether one takes the apparent diameter of M. Lescarbault's celestial object or the brightness of the objects seen by MM. Watson and Swift, we are led to think that it would need a very large number of such bodies to produce the desired effect on Mercury.
4. It is once more advisable to return to the idea first given by Leverrier, that is that there is a ring of asteroids between Mercury and the Sun; the theoretical reasons which tell in favour of the existence of this ring have lost none of their force. Leverrier had added that perhaps several of the asteroids would be big enough to be observed transiting the Sun or during a total eclipse; that was the hypothesis, it appears impossible to make a definitive judgement from the observations made up to now. (Tisserand (1882) p. 770f.).

Tisserand's lucid arguments marked the end of the first phase in the history of the anomalous advance in the perihelion of Mercury. Soon attention would be shifted to America and a new determination of the advance. Much debate of little scientific value had taken place over the question of Vulcan. Much of it obscured the merits of Le Verrier's work. He had produced a successful theory of Mercury's motion that was used in the main national ephemerides, and had isolated a fault in part of the theory. He had then put forward the hypothesis of intra-Mercurial matter as the explanation. This idea had more theoretical support than had his hypothesis of a planet disturbing Uranus which led to the discovery of Neptune. There had already been well-publicized ideas of intra-Mercurial planets before the perihelion advance had even been discovered. Bode's law, which had been used in finding new planets, did not give any obvious indication of a planet inside Mercury's orbit, though B.G. Jenkins (1878) tried to force such an interpretation. Kirkwood's analogy indicated such matter to some extent, which might explain the popularity of Vulcan spotting in the United States. Kirkwood himself joined in the debates (Fontenrose, p. 148).

Observational evidence for Vulcan was always poor, and made the matter ring hypothesis more probable than the single planet hypothesis. But the incessant search for Vulcan overshadowed the fact that even if it had been discovered, 16 similar planets would have been needed to give the 38″ perihelion advance of Mercury. The matter ring hypothesis was eventually to come into favour, and even was to be given observational support, but before that occurred other and newer ideas were to be tried.

3

NEWCOMB AND HALL'S HYPOTHESIS

3.1. Simon Newcomb

A theory such as Le Verrier's 1859 theory of the motion of Mercury was produced in order to construct tables of the motions of the planets for the use of astronomers and navigators. Mathematicians based them on the Newtonian theory of gravitation. This gave simple expressions for the gravitational interaction between two bodies, but much mathematical work was needed to produce results for a whole planetary system. The fact that the mathematics was complicated and lengthy meant that many planetary theories were found to be incorrect; the results did not agree with the planetary motions because the mathematics was faulty or incomplete. The discarded theories dented the reputations of their authors rather than that of Newton, whose gravitational scheme reigned almost unchallenged. Only a highly reputable mathematician would have had sufficient stature for his results to stand up against the authority of Newton. Such men were entrusted the task of producing theories of planetary motions to be used in the national astronomical ephemerides. These publications give the positions of the major bodies in the solar system throughout a particular year together with other relevant data. The theories used in an ephemeris have to be as accurate as possible, for the tables are constantly in use and a matter of life or death for the navigator who has calculated his position from them.

It was the need of the navigators which prompted the publication of the first ephemerides in the middle of the eighteenth century. The British *Nautical Almanac and Astronomical Ephemeris* was introduced for the year 1767, almost as soon as it had become practicable to measure a position at sea accurately using the lunar distance method (see Howse pp. 60–67). The United States of America used the British almanac until 1852, when the *American Ephemeris and Nautical Almanac* was introduced. In 1960 repetition of work was minimized with these almanacs merging into one publication (but retaining separate titles). The two almanacs used the best theories as the basis for their tables. When Le Verrier's 1859 theory of the motion of Mercury appeared it was the most advanced theory for that planet. It replaced Lindenau's 1813 theory in the 1864 volume of the *Nautical Almanac* and was used until the 1901 volume. The *American Ephemeris* used the tables of Mercury's motion prepared by Winlock that were based on Le Verrier's 1845

theory, and these lasted until the 1899 volume. In the 1900 and subsequent volumes of the *American Ephemeris* and in the 1901 and subsequent volumes of the *Nautical Almanac* Le Verrier's theories for Mercury were replaced by the 1895 theory and tables of the American mathematical astronomer Simon Newcomb.

Newcomb was a worthy successor to Le Verrier. He was born in 1835 and was largely self-educated. In 1857 he managed to obtain a post as an astronomical computer in the Nautical Almanac Office at Cambridge, Massachusetts, taking time to study mathematics under Benjamin Peirce at Harvard where he graduated in 1858. Newcomb joined the US Naval Observatory in 1861, and then in 1877 he became Superintendent of the Nautical Almanac Office, which by this time had moved to Washington. In 1884 he was made a professor at Johns Hopkins University. Newcomb died in 1909 and was buried as a rear admiral (retired) with full military honours in Arlington Cemetery.

Simon Newcomb dominated the second stage of the history of the anomalous advance in the perihelion of Mercury, in which the advance was confirmed and alternatives to Le Verrier's Vulcan sought. He backed away from solutions along the lines of Vulcan, and adopted a more unconventional explanation—the alteration of the Newtonian law known as Hall's hypothesis. Newcomb even overlapped into the third stage of the history of the Mercury advance, by giving his eventual support to a comprehensive solution which was widely accepted up to the time of Einstein. Some 30 years earlier his task had been to investigate Le Verrier's 1859 theory of Mercury's motion and to test it against newer observations. His results from this work were published in 1882.

3.2. Newcomb in 1882

Newcomb (1882) was not only concerned with the perihelion of Mercury. He also wished to find out whether observations of Mercury could be used to test for changes in the rate of the Earth's rotation in connection with tidal effects and the secular acceleration of the Moon's mean motion. Newcomb also took the opportunity of discussing the problems of observing transit phenomena for the purpose of interpreting observations of transits of Venus, which were important for the measurement of solar parallax. He used only transit observations of Mercury in his investigation of Le Verrier's theory, but used more of them (a series from 1677 to 1881) and included both internal and external contacts as his data. As always in a theory of the motion of Mercury, the mass of Venus played an important part in Newcomb's work. He compared five determinations of its value, taking the mass of the Sun as unity (Newcomb 1882, p. 467):

1. 1/347 800: from the secular motion of the perihelion of Mercury;

2. 1/408 400: from the secular motion of the node of Mercury;
3. 1/427 240: from the secular motion of the ecliptic measured by the secular motion of the node of Venus on the ecliptic;
4. no certain result: from the secular diminution of the obliquity of the ecliptic;
5. 1/401 000 or 1/396 000: from the periodic perturbations of Mercury and the Earth produced by Venus.

The values from determinations (3) and (4) were rejected as being obtained from unreliable methods. The exact value of the mass of Venus from determination (5) depended on a coefficient that Newcomb had introduced to try and ascertain whether the Earth's rotation was variable. Minimizing the residuals between theory and observation gave some indication of a variation, but little improvement was gained by assuming no variation. Newcomb evaluated the mass of Venus on this last assumption, giving the mass as 1/396 000. The mass of Venus found in determination (1) was that value which accounted for all the advance in the perihelion of Mercury, the masses of the other planets being taken as standard, and was obviously substantially greater than that in determinations (2) or (5):

There is, therefore, a decided preponderance of evidence that the true value of the mass of Venus does not differ much from 1/405 000, and is probably contained between the limits 1/400 000 and 1/410 000. The value 1/347 800 is entirely inconsistent with all the others. We must, therefore, conclude that the discordance between the observed and theoretical motions of the perihelion of Mercury, first pointed out by Le Verrier, really exists, and is indeed larger than he supposed. (Newcomb 1882, p. 472)

Newcomb had earlier commented that the validity of this reasoning was strengthened by there being here a mixture of secular and periodic determinations. Whereas all the secular determinations (1)–(4) could have been disturbed by the

continuous action of some unknown cause it was beyond all moral probability that any unknown cause should produce periodic inequalities in the planetary motions corresponding to those produced by the action of the planets on each other. (Newcomb 1882, p. 467)

In order to evaluate the anomalous perihelion advance Newcomb, similarly to Le Verrier, had to make assumptions about the other elements. The first determination of the mass of Venus was in fact dependent on both eccentricity and perihelion variations:

In investigating the actual amount of the discordance we call to mind that we have no certain evidence as to how the discordance is to be divided among the several elements which enter into the expressions for V' and W' [perturbing functions]. But, so far as has yet been noticed, it does not appear that any other element than the perihelion of Mercury is affected by this abnormal variation. We, therefore, put the inquiry into this form: assuming that the variations of e, e', and p' [primes refer to Venus] correspond to theory, how much is the variation of p [perihelion] in excess of the value given by theory? (Newcomb 1882, p. 472)

Le Verrier had obtained the variation of Mercury's eccentricity as zero by taking the orbit of the perturbing matter to be circular. Newcomb did not do this—he had as yet no explanatory hypothesis—but did not explicity state what in fact was his justification for taking the variation in e to be zero. All there was of course was the evidence from the motions due to planetary perturbations—the perihelion was already advancing 100 times as fast as the eccentricity was increasing. The advance in perihelion that Newcomb (1882, p. 473) obtained was 42″.95, slightly greater than that obtained by Le Verrier.

This paper ended with a 'Speculation on possible causes of the excess of the motion of the perihelion of Mercury' (Newcomb 1882, pp. 474–7). Newcomb first rejected the possibility of there being an error in Le Verrier's theory (this paper was of course comparing Le Verrier's theory with observation rather than putting forward a new theory of Mercury) on the grounds that the most important planet that perturbed Mercury was its neighbour Venus, and the secular variations produced by Venus had been recomputed by G.W. Hill and found to be the same as those obtained by Le Verrier. Secondly Newcomb asked whether the secular advance might be not intrinsically secular but of very long period, i.e. in so many years the advance would decrease and then turn into a secular retardation of the perihelion. However, the transit data that showed the perihelion advance spanned two centuries so that any such inequality had to have a period of at least four centuries, but suitable combinations of the periods of Mercury and Venus involved such large multiples of their mean motions that the terms would be insensible.

Having rejected these possibilities, Newcomb considered Le Verrier's hypothesis that there was intra-Mercurial matter in the form of one or a number of single planets. That this was true though 'seems to be out of the question'. Newcomb cited a number of arguments in support of this verdict and emphasized aspects of the problem that were introduced in the chapter on Le Verrier. First of all such matter had not been seen, and at the order of mass needed to produce the perihelion advance and assuming a normal relation of reflecting power to mass the planet or planets should certainly have been visible. Newcomb himself had sought these planets during the 1878 eclipse without success. It was at this eclipse that the Watson–Swift planets were discovered, 'planets' that were shown by Peters to be the stars Theta- and Zeta-Cancri. Moreover, frequent transits should have been observed. We have seen how Le Verrier eventually came to explain why so few transits had been observed. However, that was in 1876, and the degree of prediction brought in was such that the Vulcan hypothesis could be said to have been refuted. This did not touch the asteroid ring hypothesis though, and Newcomb noted the possibility of the zodiacal light being direct visible evidence for this matter. However, he thought that a collection of 100 000 bodies with a combined volume one-tenth that of the earth would glow with a much brighter light than the zodiacal light actually does . . . But we have at present no way of positively disproving it. (Newcomb 1882, p. 476)

Here we hit upon the difficulties of refutation that arose earlier in a similar connection. We shall consider the later zodiacal light hypotheses in Chapter 4, but it is worth pointing out the vagueness of Newcomb's argument rejecting this argument. It was not put on to a quantitative basis until Freundlich did so in 1915.

Newcomb had a further argument against these matter hypotheses that is new to our discussion. We have noted that in the absence of a motion in the nodes of Mercury the disturbing matter should lie in the same plane as the orbit of Mercury, i.e. at an inclinaton of 7°. We also noted that this fitted into the Vulcan hypothesis since a high inclination means less transits and more reason why these had not been observed. Newcomb had already discussed the connection between inclination and motion of the nodes and perihelion in an early paper (Newcomb 1860). Now the requirement of a high inclination for the asteroid ring seemed to be too improbable for the hypothesis to be accepted. The average inclination of the orbital plane of such a ring should eventually tend to zero owing to the perturbing influence of the other planets, i.e. the plane should lie in the plane of the ecliptic. In this position a regression of the line of Mercury's nodes would result, and this was not observed. The additional factor that a ring inclined at 7° would advance the node of Venus, the inclination of which is about half of this, can be neglected as Newcomb pointed out on the grounds that the theory of Venus was not then sufficiently sensitive to decide on this issue. This left the improbability of a ring of high inclination. However, in a letter to Bauschinger, later published in 1884, Newcomb withdrew much of the force of this argument:

I have generally objected to the hypothesis of a ring of planetoids on the ground, that they would be spread around on the invariable plane of the planetary system, and would therefore have an inclination of 5° or more to the plane of the orbit of Mercury. Then they would cause a motion of the node of Mercury, nearly equal to the motion of the perihelion.

I have since however noticed a fallacy in this statement. The theorem is true of a mean position extending through an indefinite period of time, but is not necessarily true of a great number of bodies at any one time. (Bauschinger 1884)

Newcomb then suggested that Bauschinger investigate the ring hypothesis himself. This new position was that the perihelion advance was possibly not intrinsically secular but of long period and was produced by an asteroid ring whose inclination averaged to zero over all time but was at that time about 7°. This somewhat deflated his earlier assertion that all the matter hypotheses 'must be dismissed as at least highly improbable' (Newcomb 1882, p. 476), but though the mechanical argument was now removed there still remained the basic improbability of an inclination of 7° and Newcomb did not change his opinion until forced to by weight of evidence in the early 1900s.

Newcomb also briefly considered a version of the matter hypothesis that became important later on. The line of apsides of an orbit advances if the central body is oblate. This was a standard result and had first been applied in 1758 when Walmsley had shown that the motion of the satellites of Jupiter could be explained by Jupiter's oblateness. Furthermore, according to Newtonian theory, any non-rigid spinning body is expected to be slightly oblate. Here was a possibility of explaining the advance in Mercury's perihelion by the solar oblateness. However, the Sun was not sensibly oblate, and one would need an independent reason to explain why it was apparently spherical if one wished to continue to hold that it was materially oblate. Without this, one had to conclude that the Sun was materially insufficiently oblate to produce a sensible perihelion advance.

Apart from matter hypotheses, one can explain a motion of the line of apsides of an orbit by saying that the inverse square law is not exact. Any small departure from this law produces such a motion, and Newcomb considered those departures resulting from the addition of a small extra term. Though these new laws will be considered in detail in Chapters 5 and 6, we must briefly look at Newcomb's discussion of them. The first addition was that of an inverse third or fourth power term to the inverse square term of the familiar Newtonian law. In Chapter 5 these laws are called Clairaut laws since a law of the form $I/r^2+a/r^4$ was proposed by Clairaut (1745) in order to explain the anomalous advance in the lunar perigee, where a is a constant sufficiently small to produce agreement with the inverse square law for most phenomena. Clairaut, in translating Newtonian lunar theory into modern analytical methods, discovered the anomaly which was also found by Euler and d'Alembert. Clairaut continued his investigation to try and determine the required value of the constant a, and in doing this found that in fact a fuller development of the Newtonian theory gave the correct value of the advance—the other investigations had assumed that the transverse component of the disturbing force produced no perigee advance, whereas it does though one has to go to the third power of the mass to find it. However, Newcomb had a simple objection to Clairaut laws:

> A term of the inverse third power which, at the distance of Mercury, should have a value even the millionth part of the total gravitative force of the sun would, at the distance of a foot, have a value two hundred thousand times that of the term depending on the inverse square. If higher powers than the cube were added the discrepancy would be yet more enormous. The existence of a term of such magnitude is out of the question. (Newcomb 1882, p. 472)

This seems a fairly watertight argument. There is good agreement between the relative magnitudes of the gravitational force as displayed in the long-range case of planetary motion and the short-range case of the

Cavendish experiment. We could add a small additional term of the inverse third power, and Newcomb's figures are of the right order to give the required perihelion advance, but the force between two lead spheres placed near each other would be vastly greater than that seen to twist the torsion wires gently in the Cavendish experiment. It was this experimental result that ruled out the Clairaut law.

The other type of alteration to the inverse square law that Newcomb considered in 1882 was that based on Wilhelm Weber's electrodynamic law. Weber had proposed this law in 1846 as a generalization of Coulomb's law, and it was introduced into the astronomical literature by Zöllner (1872) and Tisserand (1872). The law introduces velocity-dependent terms:

$$F = \frac{G\, m_1 m_2}{r^2}\left\{1 - \frac{1}{h^2}\left(\frac{\mathrm{d}r}{\mathrm{d}t}\right)^2 + \frac{2r}{h^2}\frac{\mathrm{d}^2 r}{\mathrm{d}t^2}\right\}$$

The constant h had been found to be equal to $\sqrt{2}\,c$ where c is the velocity of light. The only effect of using this law instead of the inverse square law is to introduce a perihelion motion of $6''.28$ for Mercury and $1''.32$ for Venus. If one did not object to using a different value of h in electrodynamics to that in celestial mechanics, one could obtain the full perihelion advance with $h = 174\,000$ km s^{-1}. Newcomb did not accept or reject this but was content to note,

> Objections have been raised to Weber's whole theory on the part of physicists, to whom the discussion of its possibility must be left. (Newcomb 1882, p. 477)

These objections will be discussed in Chapter 6.

Newcomb's discussion ended with no positive conclusions drawn. As we have seen, in 1884 the argument that he thought to be fatal to the planetoid ring form of the matter hypothesis was declared to be fallacious, and in the light of this the Newcomb rejection of matter hypotheses is not as strong as it might seem. His investigations of Le Verrier's theory of Mercury and the confirmation of a perihelion advance were now echoed in Germany. Julius Bauschinger, later to become professor and director of the observatories at Berlin, Strasbourg, and Leipzig in succession, published a monograph on the subject in 1884 together with a summary in *Astronomische Nachrichten* (Bauschinger 1884). But Newcomb returned to dominance in 1895, when he produced a work which gave the definitive value for the anomalous advance in the perihelion of Mercury.

3.3. Newcomb in 1895

In 1895 Newcomb published new tables of the four inner planets (Mercury, Venus, Earth, and Mars) that became the base for the planetary tables for the first half of the twentieth century (Newcomb 1895–8). At the same time he

published a separate discussion of the methods used and hypotheses adopted (Newcomb 1895), and it is this work, *The elements of the four inner planets and the fundamental constants of Astronomy* that will mostly concern us here. The intention remained as it was in 1882, namely that the tabular elements given by Le Verrier were being compared with the available observations. It was far greater in scope though, being for all four inner planets, using transit observations of Mercury and Venus and meridian observations for all four planets, and including values of masses determined independently of the secular variations. For each planet (except for the Earth) there were four secular variations corresponding to eccentricity of the orbit (e), longitude of the perihelion (ϖ), inclination of the orbit (i), and longitude of the ascending node (θ) which determines the orientation of the orbit. Since orbital inclinations are measured from the plane containing the Earth's orbit (the ecliptic), the latter has no inclination or longitude of ascending node, and instead values for the obliquity of the ecliptic were given. The fifteen secular variations obtained gave the amounts by which the calculated orbital elements differed from the observed values over a period of 100 years. They are shown in Table 3.1. Of the fifteen values, four were greater than their probable error. These four were therefore incorrect and errors in Newtonian theory. Four anomalies had been found in the motions of the four inner planets.
The discordances were as follows†:

1. The motion of the perihelion of Mercury. The discordance in the secular motion of this element is well known.
2. The motion of the node of Venus. Here the discordance is more than five times its probable error.
3. The perihelion of Mars. Here the discordance is three times its probable error.
4. The eccentricity of Mercury. The discordance is more than twice its probable error. It is to be remarked, however, that the probable error of this quantity is very largely a matter of judgement, and that its value may have been underestimated. (Newcomb 1895, p. 110)

The familiar advance in the perihelion of Mercury now had a companion, the perihelion of Mars. It has already been shown that as far as the perihelion is concerned the variation in longitude is obtained as a product of the variation of the perihelion with the eccentricity. If then a planet has a perihelion advance but a small eccentricity, the advance in apparent longitude is going to be small as well and quite possibly within the limits of probable error. Now the eccentricity of Venus is very small, so that the fact that its perihelion did not appear to be advancing unduly did not preclude it from actually doing so. The same is true of the Earth, so that the possibility arose of putting forward a hypothesis that gave a perihelion advance to all the four planets despite the fact that only two appeared to have such a motion.

† Newcomb refers to probable errors here, not the mean errors that appear in his Table II. A probable error is the product of the mean error and 0.674 54, as Newcomb mentioned on p. 110 of his 1895 work.

TABLE 3.1

	Observation (arcsec)	Theory (arcsec)	Difference (arcsec)
		Mercury	
$D_t e$	$+3.36 \pm 0.50$	$+4.24 \pm 0.01$	-0.88 ± 0.50
$eD_t\varpi$	$+118.24 \pm 0.40$	$\pm 109.76 \pm 0.16$	$+8.48 \pm 0.43$
$D_t i$	$+7.14 \pm 0.80$	$+6.76 \pm 0.01$	$+0.38 \pm 0.80$
$\sin iD_t\theta$	-91.89 ± 0.45	-92.50 ± 0.16	$+0.61 \pm 0.52$
		Venus	
$D_t e$	-9.46 ± 0.20	-9.67 ± 0.24	$+0.21 \pm 0.31$
$eD_t\varpi$	$+0.29 \pm 0.20$	$+0.34 \pm 0.15$	-0.05 ± 0.25
$D_t i$	$+3.87 \pm 0.30$	$+3.49 \pm 0.14$	$+0.38 \pm 0.33$
$\sin iD_t\theta$	-105.40 ± 0.12	-106.00 ± 0.12	$+0.60 \pm 0.17$
		Earth	
$D_t e$	-8.55 ± 0.09	-8.57 ± 0.04	$+0.02 \pm 0.10$
$eD_t\varpi$	$+19.48 \pm 0.12$	$+19.38 \pm 0.05$	$+0.10 \pm 0.13$
$D_t\varepsilon$	-47.11 ± 0.23	-46.89 ± 0.09	-0.22 ± 0.27
		Mars	
$D_t e$	$+19.00 \pm 0.27$	$+18.71 \pm 0.01$	$+0.29 \pm 0.27$
$eD_t\varpi$	$+149.55 \pm 0.35$	$+148.80 \pm 0.04$	$+0.75 \pm 0.35$
$D_t i$	-2.26 ± 0.20	-2.25 ± 0.04	-0.01 ± 0.20
$\sin iD_t\theta$	-72.60 ± 0.20	-72.63 ± 0.09	$+0.03 \pm 0.22$

The mean errors assigned to the theoretical values are those which result from the probable mean errors of the respective masses. They are therefore not to be regarded as independent. The mean errors given in the column of differences are those which result from a combination of those of the other two columns. The errors of the observed quantities must not, however, be judged from those of the differences, because subsequent changes in the masses of Mercury, Venus, and the Earth may produce a general diminution in the discordances. From Newcomb 1895, pp. 108–9.

The second largest discordance was that in the node of Venus. It played a central part in the ensuing discussion, but might not be entirely unexpected here for we have seen that one of consequences of the planetoid ring was to give a motion to the nodes of Venus.

Newcomb now had a complete description of the motions of the four inner planets. He knew what he had to explain, and he knew which elements were satisfactory. As before, the choice of explanatory hypotheses for the several discordances was between matter hypotheses and laws of attraction slightly different from the inverse square law. The first to be discussed were the matter hypotheses.

There was now new data relevant to the hypothesis of the non-sphericity of the Sun. Auwers (1891) had published the results of measurements of the

Sun's diameter which had been made using heliometers at five German observatories in connection with the 1874 and 1882 transits of Venus. Auwers' paper gave as the Sun's diameter $1919''.26 \pm 0''.10$, a value slightly less than previous values (1923″) which has since become the standard. However, Auwers' result showed that strictly the Sun was slightly prolate, the polar diameter being greater than the equatorial diameter by a quantity $0''.038 \pm 0''.023$. This value was of course less than the probable error in the measurement of the diameter. It was also completely opposed to what one would expect, and rather than citing it as somewhat late supporting evidence for Cartesian physics Auwers explained it as follows:

All this means nothing other than that a tendency has prevailed among the observers to overmeasure in the vertical direction compared to the horizontal. (Auwers 1891, col. 370)

The Sun then was taken to be perfectly spherical. The possibility of solar oblateness has recently become important again in connection with Dicke's modification of General Relativity which requires that General Relativity by itself does not give the exact advance in the perihelion of Mercury. If the Sun were oblate part of the 43″ advance would be explained, and the 43″ of General Relativity would introduce a discrepancy. This discrepancy could then be explained by adopting Dicke's scalar modification. Now Dicke and Goldenberg (1967) found an oblateness sufficient to cause a 3″ perihelion advance. However, this result is not uncontentious for reasons quite apart from experimental ones. The rate of rotation of the Sun about its own axis is not sufficient to cause a sensible oblateness, so that Dicke's theory requires the core of the Sun to rotate at a much faster rate than the surface and this is unstable. Newcomb had known that the oblateness due to spin was negligible, but he was not playing with an empty hypothesis. He thought it quite possible that there were vortices in the solar interior and that if the axes of rotation of these vortices (or some of them) coincided with the axis of the Sun's rotation a non-spherical distribution of matter would result. If the surface of the Sun is assumed to be an equipotential surface, the non-spherical interior should result in an elliptical surface. No one denies this, but it has been denied that a visual oblateness implies a gravitational oblateness. Ingersoll and Spiegel (1971) proposed that temperature differences between the poles and the equator could make a spherical Sun appear slightly oblate so that the Dicke–Goldenberg measurement could be accepted and still not be taken as indicating a materially oblate Sun. Dicke's conclusions are then improbable, in implying an unstable Sun, and may be fairly easily explained in alternative ways. They will be discussed again at the end of Chapter 7.

The second hypothesis to be considered was that of an intra-Mercurial ring or group of planetoids. Following Le Verrier, Newcomb assumed the ring to have zero eccentricity and produced an expression for the motion of perihelion in terms of the possible mass and distance of the ring. As before this was capable of giving an advance for Mercury but produced only 0″.031 for Mars compared with the 0″.75 required. The second objection is familiar, for the probable inclination of the ring was found to be 9°:

This great inclination seems in the highest degree improbable if not mechanically impossible, since there would be a tendency for the planes of the orbits of a ring of planets so situated to scatter themselves around a plane somewhere between that of the orbit of Mercury and that of the invariable plane of the planetary system, which is nearly the same as that of the orbit of Jupiter. Moreover, the motion of the perihelion of Mars is still unaccounted for and that of the node of Venus only partially accounted for, as shown by the large residual of the second equation. In fact, the great inclination assigned to the ring comes from the necessity of representing as far as possible the latter motion. (Newcomb 1895, p. 114).

The dynamical argument against a high inclination did not hold for a single planet but that (Vulcan) was ruled out on observational grounds. Newcomb also felt that such a ring would be visible, but he offered no quantitative argument on this so that this was a weak objection. It was also weak to object that the perihelion advance of Mars was not produced since there was nothing against combining the matter hypotheses and having, say, more than one ring. Newcomb was content to reject these hypotheses singly but one can overcome his principal objections by suitable combinations of matter hypotheses. In the present case it is the dynamical argument that is the strongest so that a ring of low inclination is not improbable. Newcomb should then have shown that the node of Venus could then not be explained and by implication that the hypothesis was refuted, but he did not do this. Though the letter to Bauschinger was referred to in the section introducing the discussion of the hypotheses, the partial retraction of the dynamical argument given in that letter was not mentioned. The possibility, that Newcomb allowed in that letter, might still be regarded as improbable.

The third of the matter hypotheses to be considered was that 'of an extended mass of diffused matter like that which reflects the zodiacal light' (Newcomb 1895, p. 115).† Newcomb had two objections to this. First the matter appeared to lie in, or at least be not much inclined to, the ecliptic, and so it would produce motions in the nodes of Venus and Mercury which were the opposite of those found. Secondly the zodiacal light matter extended beyond the orbit of the Earth and thus beyond the orbit of Mercury. Matter wholly inside an orbit and matter wholly outside an orbit would both produce an advance in perihelion, but matter in that part of the orbit enclosed between the aphelion and perihelion distances would produce a retrograde motion of the perihelion. If it is assumed that the density of the diffuse matter decreases as the distance from the Sun increases, the effect of the matter outside the aphelion will be largely cancelled by the matter within the aphelion/perihelion ring. The

† The nature of the zodiacal light is discussed in Chapter 4.

effective matter is then the intra-Mercurial matter, the objections to which have already been outlined. However, Newcomb was wrong to infer that those objections worked against this hypothesis. They were in brief that the perihelion of Mars was not represented, that the inclination was improbable, and that the matter ought to be visible. Here though the matter was visible and it appeared to lie in the ecliptic. The only objection that carried over was that Mars was still discordant, but it has already been pointed out that it was still possible that another ring might be brought in to explain this. We are thus left with the objection that, because this matter appeared to lie in the plane of the ecliptic, it would produce a retrograde motion in the nodes of Venus and Mercury instead of the observed advance. The problem with the zodiacal light matter was to determine its mass. We shall see in Chapter 4 that later workers assigned it a substantial mass. But Newcomb assigned it negligible mass. It is important because, since few deny that the matter is there, almost everyone accepts that there is disturbing matter in the solar system that is not taken into consideration when theories of planetary motion are being constructed. There are two ways of determining the mass of planets having no satellites. One can look at the perturbative effect of the planet on another body, such as when Encke's Comet was used to measure the mass of Mercury, or one can look at the brightness and compare it with plausible albedo assumptions, as was done for Pluto. Either of these methods can be used in the case of the zodiacal light. In the first case it can be held responsible for the perihelion advance of Mercury and be assigned a large mass. In the second place assumptions can be made about the brightness of small particles and comparisons made with the brightness of the zodiacal light and a small mass will be given. The assumptions in the second method are such that they can also be used to obtain agreement with the value in the first method. Now at this point Newcomb did neither of these things but, because he thought the matter would not explain his discordances, inferred that the mass is small.

The fourth and final matter hypothesis that was considered by Newcomb was that 'of a ring of planetoids between the orbits of Mercury and Venus' (Newcomb 1895, p. 116). He gave a complete specification of such a ring sufficient to account for the anomalies in Mercury and Venus but found two objections to it. First was the familiar one of the improbability of a ring of high inclination—in this case 7°.5. This was less than the 9° previously objected to, but was just greater than that of the plane of Mercury and produced the required advance in the nodes. Whereas before the 9° had been 'in the highest degree improbable if not mechanically impossible' (Newcomb 1895, p. 114), the tone was now a little lower:

> In admitting such orbits we encounter difficulties which, if not absolutely insurmountable, yet tell against the probability of the hypothesis. (Newcomb 1895, p. 117)

The next objection though was new, and concerned the possibilty of explaining the advance in the perihelion of Mars by the action of the minor planets. This involved Newcomb in a dilemma. Either their total mass was too small to produce any appreciable effect, or if it was sufficient it would mean a large number of small minor planets, so many as to produce a 'zone of light across the heavens'. The argument was a probabilistic one based on the number of minor planets having a given magnitude (which is a function of mass). However, as he pointed out, there was at least some light where Newcomb needed it—the matter producing the gegenschein and a faint continuation of the zodiacal light—but 'their total mass [is] too small to produce any appreciable effect'. What was needed was a compromise—a sufficiently large mass that was in such a form as to be no brighter than the faint zone seen. This

> is a very important question which cannot be decided without exact photometric investigations. It is, however, certain that if we could do so we should have to suppose a very unlikely discontinuity in the law of progression between each magnitude and the number of bodies having that magnitude'. (Newcomb 1895, p. 118)

The hypothesis was then not completely knocked out but was improbable. We may now list the four matter hypotheses and those objections of Newcomb that seem strongest.

1. Non-sphericity of the Sun: Auwers' results;
2. Intra-Mercurial ring: high inclination to account for node of Venus;
3. Zodiacal light matter: gives retrograde motion to nodes of Mercury and Venus;
4. Ring between Mercury and Venus: high inclination.

3.4. Hall's hypothesis

The difficulties that Newcomb found with the matter hypotheses were not insuperable. The alternative approach—changing the inverse square law—will now be considered to see how effective that was.

In Chapter 1 a theorem of Newton was introduced that gave the advance in the line of apsides for laws of attraction other than the inverse square law. For a law of the form

$$F = \frac{br^m - cr^n}{r^3}$$

where b, c, m, and n are constants, the angle between successive perihelia is given by

$$\theta = 2\pi \left(\frac{b-c}{mb-nc} \right)^{\frac{1}{2}}.$$

Clairaut's law is given by putting $m=1$ and $n=-1$. However, there is the possibility of putting $c=0$ and then putting $m=1+\delta$, where δ is a small number much less than unity. This would appear to approximate to the inverse square law, give a perihelion advance, and yet not be subject to Newcomb's objection against Clairaut laws. Newton himself used as an example such a force law (Newton 1934, p. 147 (Book I, Prop. XLV, Cor. II)). The application of a law of this sort to the problem of Mercury's perihelion was suggested by Asaph Hall (1894) and since he was the only person to suggest it the law may be called Hall's hypothesis. He cited as his authority Bertrand who had shown that the only central force laws of simple distance dependence that gave closed orbits were $F(r)=Ar^{-2}$ and $F(r)=Ar$ (Bertrand 1873). He had considered a law of the form

$$\phi(r) = \frac{A}{2} r^{1/m^2-3}$$

and showed that only for $m=1$ or $m=\frac{1}{2}$ was the orbit closed. The perihelion advanced an amount $2m\pi$ in a revolution, which gave closed orbits for $m=1$ and $m=\frac{1}{2}$. Hall considered a central force of the form $F=c\cdot r^n$, for some constant c, and by following Bertrand's result found that the perihelion advance was given by $\theta=2\pi\cdot(n+3)^{-\frac{1}{2}}$. The orbit was closed for the inverse square case. Hall thus found that the perihelion advance of Mercury, 43″ per century, could be represented by putting $n=-2.000\,000\,16$, multiplying by the mean motion, and adjusting the units. Newcomb took the perihelion advance to be 42″.34 and found $n=-2.000\,000\,1574$ (Newcomb 1895, p. 118).

Given that a change was needed, Newcomb found this law to be preferable to Weber-type laws:

This hypothesis seems to me much more simple and unobjectionable than those which suppose the force to be a more or less complicated function of the relative velocity of the bodies. (Newcomb 1895, p. 118)

Hall's hypothesis, with n chosen to fit Mercury in an *ad hoc* way, was found by Newcomb to represent the other planets as well (Table 3.2). In fact agreement

TABLE 3.2

Planet	Discordance ($D_t\varpi$) (arcsec)	Hall (arcsec)	Discordance ($e\cdot D_t\varpi$) (arcsec)	Hall (arcsec)
Mercury	+42.4	+42.34	+8.48±0.43	+8.70
Venus	−7.0	+16.58	−0.05±0.25	+0.11
Earth	+5.9	+10.20	+0.10±0.13	+0.17
Mars	+8.3	+5.42	+0.75±0.35	+0.51

It will be seen that the evidence in the case of Venus and Earth is negative, owing to the very small eccentricities of their orbits, while the observed motion in the case of Mars is very closely represented. (Newcomb 1895, p. 119)

within probable errors in $e \cdot D_t\varpi$ was found for all four planets. The table for $D_t\varpi$ shows a general difficulty for the hypothesis. The Hall perihelion motions decrease for increasing radius of orbits (i.e. for decreasing mean motions since the perihelion shift in each orbit is the same for all planets and satellites on Hall's hypothesis) whereas the values of the discordances show no such regularity. Indeed the perihelion of Venus appears to regress rather than to advance, albeit with an error that allows for an advancing value. This seems a prima facie reason for rejecting any hypothesis that attempts to explain these discordances solely in terms of advances, but this is being too simplistic. What is observed is not $d\varpi$ but $e \cdot d\varpi$, so that the discrepancy value of $D_t\varpi$, if it is dominated by observation, might be expected to follow the general drift of the $e \cdot D_t\varpi$ values as observed. This is in fact what is found. The values of $e \cdot D_t\varpi$ (discrepancy and Hall) and $D_t\varpi$ (discrepancy) all show a large value for Mercury, and then a low value for Venus increasing to the value for Mars.

Two problems now faced Newcomb. Was there independent evidence for Hall's hypothesis in the case of satellites? Could the remaining anomalies, of which the advance in the nodes of Venus was most important, be explained in other ways?

3.5. A lunar digression

The answer to the first question involves a short digression into a particular controversy in nineteenth century lunar theory. It was only in the case of the Moon that theory was sufficiently advanced to give any hope of answering whether there was an anomalous advance in the line of apsides, an advance that according to Hall's hypothesis should have been about $1''.4$ per year:

This is very nearly the hundred-thousandth part of the total motion of the perigee. The theoretical motion has not yet been computed with quite this degree of precision. The only determination which aims at it is that made by Hansen. He finds

	Theory ($''$)	Obser.($''$)	Diff.($''$)
Annual mot. of perigee.	146434.04	146435.60	+1.56
Annual mot. of node	−69676.76	−69679.62	−2.86

The observed excess of motion agrees well with the hypothesis, but loses all sustaining force from the disagreement in the case of the node. The differences Hansen attributes (wrongly, I think) to the deviation of the figure of the Moon from mechanical sphericity. (Newcomb 1895, p. 119)

Now the motion of the perigee seemed to confirm Hall's hypothesis. The value of the small number δ had been determined *ad hoc* to fit the perihelion of Mercury, it was then found to agree with the other three inner planets, and now it agreed with the motion of the lunar perigee. However, Newcomb did not take it this strongly, for there was still the lunar node which Hall's hypothesis predicted as stationary. Therefore another cause had to be found which produced this motion, but any such cause was likely to have other effects: a perturbative cause would also produce a motion of the perigee and destroy the agreement with Hall. Suggesting that the motion of the node was due to an error in constructing the theory raised the possibility that there was also an error in the perigee motion, and we shall see later that Newcomb refused to take this position in a similar situation even though by not doing so he involved himself in improbabilities. Here the possibility of error was not mentioned. The position that he did adopt was that the theory was not sufficiently accurate to decide the issue, so that Hall's hypothesis was neither confirmed nor refuted.

This position was the exact opposite of that originally taken by Hansen, for he had used these values to determine the figure of the Moon. According to Hansen there were no anomalous motions of the perigee or node because the apparent differences were attributed to the Moon's figure being not exactly spherical. This also included the possibility that the Moon was spherical in shape but that the centre of gravity and the centre of the figure did not coincide, and it was none other than Newcomb who had attacked this in 1868 on the grounds of inconsistency (see below).

Hansen's lunar tables had been published in 1857. The most accurate of their type, they became the standard tables for the ephemerides. They were lavishly praised by the Astronomer Royal, Airy, and Newcomb himself described Hansen as the 'greatest living master of celestial mechanics since Laplace' (Newcomb 1903, p. 315). The theory behind Hansen's tables was not published until 1864, but a paper had already been published in 1856 entitled *Sur la figure de la lune* (Hansen 1856). Hansen found that in order to obtain agreement with observation the theory required that the centre of the figure of the Moon did not coincide with the centre of gravity. This possibility was quite new, and Hansen was able to show that non-coincidence of the centres required the inequalities in the mean longitude of the Moon to be multiplied by a constant factor depending on the distance between the two centres. The basic idea is seen in Fig. 3.1. In its mean position the Moon is at B and, to the observer, the centre of gravity c_g and the centre of the figure c_f coincide. Owing to the inequalities in the Moon's motion it sometimes appears ahead of the mean position (at C) and sometimes behind it (at A). The observer measures the angle between these positions as the angle θ_f, but the theoretician, who uses the centre c_g, computes an angle θ_g. Since $\theta_g < \theta_f$ the theoretical inequalities appear to be less than the inequalities observed, and to obtain agreement the former must be multiplied by a factor slightly greater than unity which is a function of the distance $c_g - c_f$. The chief inequalities are the evection, variation, and the annual equation, and using observations made at Greenwich and at Dorpat Hansen found that each needed to be increased. Airy had come to the same conclusion when he compared Plana's theory with

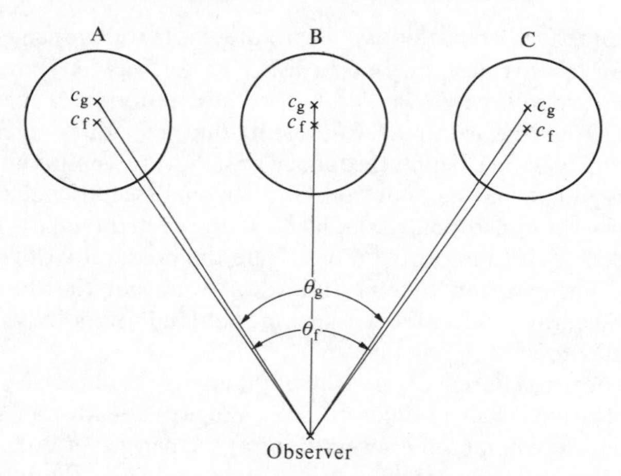

FIG. 3.1 Observing Hansen's Moon.

a series of observations at Greenwich (Airy 1849). Plana's evection of 4585″.6 had to be augmented by 1″.28, and Hansen's evection of 4585″.94 had to be augmented by 0″.69. The correction factor was found to be 1.000 1544, which corresponded to a separation of the centres of 59 000 m. Though no probable errors were given, Hansen thought that his theory was sufficiently accurate to support this conclusion.

It followed from this that the gravity at the surface of the Moon was greater on the far side than on the side facing us, so that, whereas we saw a sterile region, 'we may not conclude that the other hemisphere is not endowed with an atmosphere, and that it has no vegetation and living beings' (Hansen 1856, p. 32). The possibilities of extra-terrestrial life, and the problems attendant on it, were the subject of much debate in the nineteenth century and Hansen's theory was used to support speculations on lunar inhabitants (Proctor 1873, p. 221, Newcomb 1903, p. 316).† However, it could also be used to answer a more mundane question: What had happened to the lunar atmosphere? Though it was not totally accepted that there was no atmosphere visible on the side facing us (Harley 1885), there were four common answers available to those who accepted scientific opinion that there was none. Whiston thought that a comet could have removed the oceans and atmosphere in a near collision. Hansen's hypothesis was used to suggest that the atmosphere and oceans had been attracted round to the far side of the Moon, there to remain unseen. Others thought the oceans and air had passed into cavities in the lunar surface or that they had become frozen. Proctor (1891) thought that the third suggestion was least unlikely, citing as his reasons against the Hansen

† Both Proctor and Newcomb refer to such speculations but neither gives names to the speculators.

hypothesis Newcomb's 1868 objection, which will be considered below, and also that libration allowed us to see part of the far side and still no atmosphere was observed.

In his major work of 1864, Hansen approached the problem from a different angle. This was the work referred to by Newcomb in his 1895 discussion. Hansen gave expressions for the excess in the motions of the node and perigee in terms of the differences between the three moments of inertia A, B, and C. Using the observed excesses in these motions, i.e. the values later cited by Newcomb in 1895, he found the value of a factor f, where $f=(C-B)/(C-A)$, calculating it to be 0.9. Now Laplace had obtained a value of 0.25 by considering the Moon as a fluid body, Nicollet had obtained 0.056 from observations of libration, and later Wichmann had found 0.419 using the same phenomenon (Harkness 1891, p. 102). Hansen's value was much higher than these, though the three values obtained before his could hardly be said to have agreed. This lack of agreement and the weakness of the determinations allowed Hansen to uphold his own value without involving too much improbability, though one might still charge Hansen's own value with weakness. It was as strong as his theory, which Hansen thought was quite strong enough. Laplace's value was as strong as his assumption, which concerned the Moon cooling from a fluid state while in motion around the Earth and which one would not take as being of great strength.

In 1868 Newcomb attacked Hansen's hypothesis of 1856 regarding the figure of the Moon (Newcomb 1868). He was supported by Delaunay (1870), who described Newcomb's chief argument as a 'capital objection'. Hansen (1871) replied, poking fun at Newcomb's assertion that the hypothesis was 'devoid of logical foundation' but not really replying to the objection. Newcomb had argued that on Hansen's theory the theoretical evection was not less than the observed evection but was in fact equal to it:

Let e be the true eccentricity of the orbit described by the moon's center of gravity. Then the true evection in the same orbit will be $e \times A$; A being a factor depending principally on the mean motions of the sun and moon. And on Hansen's hypothesis, the *apparent* evection, or that of the center of figure, will be $e \times A \times 1.000\,1544$.

On the same hypothesis, the eccentricity derived from observation, being half the coefficient of the principal term of the equation of the center, will be $e \times 1.000\,1544$, and the theoretical evection computed with this eccentricity will be $e \times 1.000\,1544 \times A$, which is the same with that derived from observation. Hence, the theoretical evection will agree with that of observation, notwithstanding a separation of the centers of gravity and figure of the moon. (Newcomb 1868, p. 377)

This argument is very forceful. One cannot adjust theoretical quantities to fit observational quantities by introducing a 'conversion' factor if the theoretical quantities contain terms that have been determined by observation. The second part of Newcomb's attack on Hansen's theory was this: out of the inequalities which escaped the first argument, only the variation was sufficiently large to be used in judging a discrepancy between theory and observation. As far as this was concerned, Newcomb regarded the observational value as too imprecise and a provisional value agreed fairly well with the theoretical value. The evidence for Hansen's 1856 theory was thus nil.

Now Newcomb's 1895 remark that he thought Hansen wrong in holding that the Moon was not mechanically spherical referred to the 1864 theory. However improbable it might appear, one can at least conceive of a spherical Moon whose centre of gravity was not coincident with the centre of the sphere. That the Moon is not spherical is quite probable. It is a spinning body so one would expect oblateness. The same face is always turned towards the Earth so one would expect some 'tidal' bulge in that direction. The problem had been whether the differences between the three moments of inertia were sufficient to produce large motions of the perigee and node. However, the problem facing Newcomb was the opposite of this: Were the perigee and node motions sufficiently well determined to allow it to be said that the Moon was more non-spherical than had previously been thought? Newcomb, having had this previous experience of Hansen overstepping the boundaries of accuracy, decided that they were not. This had the additional consequence that the node motion could not be regarded as refuting Hall's hypothesis which predicted the perigee motion but not that of the node.

Though Newcomb did not have long to wait before a successor to the lunar theory of Hansen appeared, evidence relevant to this situation was already available since Hill (1884) had published a paper with the title *Determination of the inequalities of the Moon's motion which are produced by the figure of the Earth*. Just as an oblate Sun produces a perihelion motion, an oblate Earth produces a perigee motion and, if the lunar orbit does not coincide with the terrestial equator, a motion of the lunar node. Hill obtained $+6''.8201$ and $-6''.4128$ respectively for the annual motions of the perigee and node due to the figure of the Earth. These compared with Hansen's values of $+5''.87$ and $-5''.90$. Considering the perigee motion alone, the excess of observation over theory was then $0''.7$ for Hansen–Hill and $1''.6$ for Hansen (without the Hill modification) compared with the value of $1''.4$ predicted on Hall's hypothesis. The Hansen–Hill theory was thus incompatible with Hall's hypothesis. Furthermore, the anomalous motion of the node was decreased from the Hansen value of $-2''.7$ per year to $-2''.2$, so that if one rejected the Hansen hypothesis of the figure of the Moon the Hansen–Hill lunar theory provided a serious competitor to the Hansen–Hall theory. Now Hill's paper was not without contentious points. Brown (1904) wrote that he thought that Hill's investigation had not been carried to sufficiently high orders, but further work showed that Hill's values for the annual perigee and node motions became $+6''86$ and $-6''.42$ respectively, which is not a large change. However, Hill's value for the ellipticity of the Earth was certainly contentious. His value of 1/286.7 was obtained from pendulum observations. A comparison with other values, such as those given by Harkness (1891), shows Hill's value to be larger than the rest, even than the pendulum results of Helmert. This would give grounds for requiring Hill's ellipticity value to be decreased, and this would of course have the effect of decreasing the motions in the lunar perigee and node back towards the Hansen level.

3.6. The advance in the nodes of Venus and the solar parallax

The lunary theory was rather equivocal about the introduction of Hall's hypothesis. A few years later it was possible to make a clear decision on this question using Brown's lunar theory, but before that is considered we have still to finish Newcomb's 1895 discussion. The problem we have left was whether, having explained the perihelion anomalies using Hall's hypothesis, it was possible to explain the advance in the nodes of Venus. Rather than introduce an extra hypothesis to explain this motion, Newcomb tried to see whether he could determine a new set of planetary masses which would satisfy the remaining secular variations. This he found was quite possible:

> Speaking in a general way we may therefore say that the representation of the secular variations, those of the perihelia being ignored, by these corrections to the masses is satisfactory. Except for the large discordance in the motion of the eccentricity of Mercury the mean error would have been less than unity. (Newcomb 1895, p. 122)

This looked promising, for the combination of Hall's hypothesis and this set of masses looked as if it would clear up all the anomalies with only the eccentricity of Mercury remaining recalcitrant. However, this set of masses had to agree with independent determinations and at this point trouble appeared:

> Comparing the two sets of values we find that the masses of Mercury, Venus, and Mars agree well with those derived from other sources. Very different is it with the mass of the Earth. The discordance is here more than the hundredth part of its whole amount, which involves a discordance of more than the three-hundredth part in the value of the solar parallax. (Newcomb 1895, p. 122)

Since the solar parallax is a direct function of the mass of the Earth,† and the solar parallax had been determined many times, Newcomb had a number of values for comparison to see to what extent the new mass of the Earth was discordant. These solar parallax values are shown in Table 3.3. The theory behind each determination will now be briefly given.

Determination 1. This arises from having to allow for the Sun's motion around the centre of mass of the Sun–Earth system. It can be shown that $n^2a^3 = G(M+m)$, where $n = 2\pi/T$, T is the period of the Earth's orbit, a is the mean distance, M is the mass of the Sun and m that of the Earth. We thus have a connection between m and a.

† See determination 1.

TABLE 3.3
Results of determinations of the solar parallax arranged in the order of magnitude

(I)	From the mass of the Earth resulting from the secular variations of the orbits of the four inner planets . . .	8″.759 ± 0″.010	wt. 9
(II)	From Gill's observations of Mars at Ascension . . .	8″.780 ± 0″.020	wt. 2
(III)	From Pulkowa determinations of the constant of aberration . . .	8″.793 ± 0″.0046	wt. 40
(IV)	From observations of contacts during transits of Venus . . .	8″.794 ± 0″.018	wt. 3
(V)	From the parallactic inequality of the Moon . . .	8″.794 ± 0″.007	wt. 18
(VI)	From determinations of the constant of aberration made elsewhere than at Pulkowa	8″.806 ± 0″.0056	wt. 28
(VII)	From heliometer observations on the minor planets . . .	8″.807 ± 0″.007	wt. 20
(VIII)	From the lunar equation in the motion of the Earth	8″.825 ± 0″.030	wt. 1
(IX)	From measurements of the distance of Venus from the Sun's center during transits . . .	8″.857 ± 0″.023	wt. 2

From ref. Newcomb 1895, p. 157.

Determinations 2 and 7. These use Kepler's third law $a^3/a_1^3 = T^2/T_1^2$, where the subscript 1 refers to a near exterior planet (e.g. Mars or a minor planet). The distance d is directly measured (e.g. for the same observatory the position may be measured a few hours before and after meridian passage), where d is the distance between the earth and the planet at some favourable time (e.g. opposition) and is small enough to be measured directly. The values of d and the ratio a^3/a_1^3 can then be used to find a.

The use of minor planets was considered preferable to the use of Mars on the grounds that the sensible disc of the latter made centring of the telescope threads much more difficult than for a (practically) point image. Newcomb also discussed the influence of the red colouration of Mars, which by differential refraction between it and a non-red comparison star would produce a measured parallax in excess of the true value. Nevertheless he used Gill's value, admittedly with a low weight (only 2), because Gill had found no evidence for this effect and because his value was smaller than all the other accurate measurements. Newcomb appeared slightly embarrassed by this move:

> The principle that when a result is open to a strong suspicion of being affected by a cause which would cause it to deviate in one direction, it is logical to conclude *a posteriori* that the cause has not acted if the deviation is found to be in the other direction, may not be a perfectly sound one, but I have nevertheless acted upon it. (Newcomb 1895, pp. 154–5)

Later, when modifying the weights, Newcomb halved the weight of this so he did take note of the problem. That Gill had found no evidence for the effect when comparing the observations of Mars when it was at different altitudes supported Newcomb's acceptance of the value. The second reason could be used to support the value from (1), for if (2) *was* too large it would, being larger than (1), give support to it.

The 'red-shift' was supported elsewhere, for the minor planets are of different colours, and of these Iris is reddish:

> Now, in this connection, it is a significant fact that the parallax found from observations of Iris, 8″.825, is the largest by Gill's method. (Newcomb 1895, p. 165)

Newcomb modified the weight of (7) from 20 to 5 and used just Victoria and Sappho ('I am not aware of any evidence that Victoria and Sappho differ in color from the average of the stars').

Determinations 3 and 6. The constant of aberration is given by the velocity of the Earth in its orbit divided by the velocity of light. Using the constant of aberration, the velocity of light, the period of the Earth's orbit, the eccentricity of the Earth's orbit, and the equatorial radius of the Earth we can find the solar parallax (see Smart 1965). The two values (3) and (6) came from using different values of the constant of aberration. Newcomb assigned a very high weight (40) to the Pulkowa determination which was retained, and a fairly high one (28) to the miscellaneous determinations which was later reduced to 10. The Pulkowa value was a mean of 10 values made by different observers using different methods. The value obtained and the 'miscellaneous' value differed by more than the sum of their probable error—this left 'room for a suspicion of constant error in one or both means' (Newcomb 1895, p. 139). There are in fact many possible errors, both systematic and accidental, in these determinations. One might cite the then recent discovery by Chandler of the variation in latitude which had to be allowed for.

The weights were initially assigned according to the rule that a mean error of $\pm 0''.30$ corresponded to a weight of 1,

> allowance being made, however, for doubt as to what value should be assigned to the mean error and for the different liabilities to systematic error. (Newcomb 1895, pp. 156–7)

Now the high weights came partially from the latter, for the two determinations were each composed from results obtained from different observers with different instruments and using various methods, and as well

> It is not to be supposed that any of the systematic errors already indicated would pertain to all observers and to all instruments. (Newcomb 1895, p. 134)

What does seem questionable was the differential treatment when the Pulkowa value retained its full weight and the 'miscellaneous' value had its weight more than halved. The final value of the solar parallax obtained by Newcomb was utterly dominated by the Pulkowa value.

Determination 4. The transits of Venus used to be the standard and best way of measuring the solar parallax. This determination used the transits of 1874 and 1882 and the earlier famous ones of 1761 and 1769. The mean for the whole series (16 in all) was $P_1 = 8''.797 \pm 0''.023$ (weights assigned according to mean errors). Separating internal and external contacts gave $P_2 = 8''.796 \pm 0''.023$ and $P_3 = 8''.908 \pm 0''.06$ respectively, indicating that less weight should be given to the latter. This modified mean was $P_4 = 8''791 \pm 0''.022$. The agreement of errors in the two means showed that both were equally probable so that a final value was obtained by combining them with equal weights to give $P = 8.794'' \pm 0.018''$.

Determination 9. This was a variant of the transit determination (4), using heliometers and photoheliographs, for the 1874 and 1882 transits. In terms of weights the determinations were dominated by a photographic determination by Harkness in 1882, and Newcomb drastically reduced this weight from 54 to 6. The mean error in the tabular result for P was obtained by summing 'all known sources of error' and increasing in proportion of the total weights (29) to the heliometer weights (15). This error, $0''.023$, was larger than the error of $0''.016$ obtained by comparing the mean with the individual values. Newcomb was not in favour of the method, for it

> is therefore subject to this serious drawback: that the parallax depends on the measured difference between arcs [centre of Venus to centre of Sun] which may be from thirty to fifty times as great as the parallax itself, the measures being made in different parts of the Earth. (Newcomb 1895, p. 143)

Newcomb assigned the low weight of 2 to this value which was later reduced to 1 (the weight of (4) is 3).

Determination 8. The lunar inequality arises in the Earth's motion due to the terrestial motion around the centre of mass of the Earth and Moon. This motion is affected by the Sun's perturbation on the Moon, and an expression can be derived connecting the solar parallax, the mass of the Moon, and the lunar inequality. Newcomb found $P = 8''.818 \pm 0''.030$, a value obtained by using for the coefficient of the lunar inequality a mean of two values from separate observations by Gill and Newcomb himself. The large error in P was reflected in its weight of 1. However, it was retained because

> I do not think it is liable to systematic error, and we must therefore regard the mean error assigned as real. (Newcomb 1895, p. 166)

The weight remained as 1.

Determination 5. The parallactic inequality in the Moon's motion is due to the disturbance of the Moon by the Sun being greater when they are nearer and less when they are further away from each other, and is thus a function of the solar parallax. Newcomb used the mean of three determinations, his own value from meridian observations at Greenwich and Washington, Battermann's value from occultations, and Franz's value from observations of the lunar crater Mösting A, the respective weights being 5:2:1. His own work involved determining a small correction to a solar parallax of $8''.848$, the corrections being hopelessly discordant amongst themselves. The final mean was $P = 8''.794 \pm 0''.008$, which was given a weight of 18. In the final adjustment Newcomb admitted to uncertainty about the errors, but all this meant was that

> While it is true that the value may be smaller than what we have assigned, we must also admit that it may be much larger. (Newcomb 1895, p. 165)

However, the weight was reduced to 10.

This account of Newcomb's discussion of the solar parallax values has been given because of the central role that they will now play. We have seen that the value obtained using the mass of the Earth seemed to be discordant, and this can be used to test the efficacy of explaining the motion of the nodes of Venus by decreasing the mass of the Earth. The motion of the node was measured on the ecliptic which was itself moving. The later motion depended mainly on the mass of Venus, and this had already been checked because of its effect on the perihelion of Mercury. Since therefore the mass of Venus could not be changed, the motion of the ecliptic could not be changed, so that the anomalous motion of the node on the ecliptic had to be due to the node itself. Now the perturbation of Venus by the Earth caused a regression of the nodes as seen on the ecliptic, and the more massive the Earth the greater the regression. The advance in the nodes of Venus which was being discussed was due to the theoretical regression being greater than the observed regression, so that instead of the node having regressed $106''$ it had only regressed $105''$. There appeared then to be an advance of $1''$. To reduce this the theoretical regression had to be reduced, which in turn meant reducing the mass largely responsible, i.e. the mass of the Earth. Now this meant, using the relationship behind method (1), reducing the solar parallax and it was fortunate that there were independent means of checking this reduction. Before the comparison with the other values Newcomb checked that the advance was well founded. He did so, and thought that it was so well founded that the clash with the other parallax values resulted in the following remarks found within the space of a few pages:

> What adds to the embarrassment and prevents us from wholly discarding the suspicion that some disturbing cause has acted on the motion of Venus, or that some theoretical error has crept into the work, is, that, of all the determinations of the solar parallax this is the one which seems most free from doubt arising from possible undiscovered sources of error. (Newcomb 1895, p. 159)

> This result of observation, errors and unknown actions aside, I can not suppose to be affected by any other mean error than that here assigned. (Newcomb 1895, p. 163)

> Unknown actions and possible defects of theory aside, it seems to me that the value of the solar parallax from this discussion is less open to doubt from any known cause than any determination that can be made. (Newcomb 1895, pp. 163–4)

> The embarrassing question which now meets us is whether we have here some unknown cause of difference, or whether the discrepancy arises from an accidental accumulation of fortuitous errors in the separate determinations. (Newcomb 1895,p. 177)

Note how in each quotation Newcomb mentioned the possibility of error in the theory. Now values of the solar parallax were being compared rather than the motion of the nodes directly, so that it was at least possible that, rather than the new mass of the Earth being wrong, some other factor used in determining the solar parallax from this mass (method 1) was at fault. However, these factors were standard and well attested. It was the advance of the node that was being tested, and the corresponding value of the solar parallax was too small. If it was correct the other determinations were all systematically too large, a result which was improbable in view of the separate nature of their evaluations. Now one could argue that independence of method is not of paramount importance and that when complicated methods are being used there are so many influential factors that the determined quantity will only be judged by its agreement with previous determinations. This 'sociological' dependence finds some support in the parallax values of the star 61 Cygni. After Bessel's prize-winning measurement of this quantity in 1838, succeeding values rose until the turn of the century and then gradually decreased until the value accepted today is close to Bessel's original value. The standard Ephemeris value of the solar parallax for 1870–1899 was Newcomb's 1867 value of $8''.848$, and this was considerably greater than most of the values under discussion. If one could prove the 'sociological' position it would mean less improbability in reducing seven determinations at once, simply because the apparent agreement of the seven values is expected and says little about the 'true' value. A more conventional position is that the simultaneous reduction is improbable because it requires all seven values to be systematically wrong in the same direction. As Newcomb saw it:

> The question thus takes the form whether it is possible that the mean of the seven determinations of the solar parallax $P = 8'',797 \pm 0.0035''$ can with reasonable probability be in error by an amount the correction of which would bring it within the range of adjustment of the other quantities. (Newcomb 1895, p. 167–8)

Now all the determinations were open to both systematic and accidental errors. One could have objected to the theory behind some of the methods, the main ones being the aberration values (3) and (7). Newcomb realized this, but

> I can only remark that its simplicity and its general accord with all optical phenomena are such that it seems to me that it should be accepted, in the absence of evidence against it (Newcomb 1895, p. 147)

We have seen that none of the solar parallax determinations gave the order of certainty that was desirable, but for each one it was equally likely that a redetermination would increase it as would decrease it. Newcomb clearly recognized this:

> The improbability which meets us is not so much the amount of the individual errors of the determinations as the fact that seven of the eight independent determinations should all be largely in error in the same direction. (Newcomb 1895, p. 168)

Nevertheless he chose to make this improbable move:

> Still, under the circumstances, we must admit this possibility, and make what seems to be the best adjustment of all the results. (Newcomb 1895, p. 168)

3.7. Newcomb and Hall's hypothesis

Before criticizing Newcomb's position, we must take stock of his assertions. It is clear that he has rejected the matter hypotheses—this was stated explicitly. However, his positive acceptances were not at all strong:

> What I finally decided on doing was to increase the theoretical motion of each perihelion by the same fraction of the mean motion, a course which will represent the observations without committing us to any hypothesis as to the cause of the excess of motion, though it accords with the result of Hall's hypothesis of the law of gravitation; to reject entirely the hypothesis of the action of unknown masses, and to adopt for the elements what we might call compromise values between those reached by the preceding adjustment and those which would exist if there is abnormal action. (Newcomb 1895, p. 174)

The values were compromise ones because they did not fully cover the non-perihelion anomalies. There was for instance the advance of the nodes of Venus, which still remained the largest anomaly at $0''.25$ (it had been $0''.60$), which was about 1.5 times its probable error.

Newcomb was also wary about Hall's hypothesis. I think that Hall's hypothesis is improbable in itself, as will be discussed later, and Newcomb might or might not have felt the same, but I do not think that the position of the last extract supports at all such claims as those of Fontenrose (1973, see especially p. 155)—that this was a departure from Newtonian physics—or of Hanson (1962)—that this was a conceptual shift and modification of foundations of Newtonian theory. One could argue that Newcomb did not believe that Hall's hypothesis really represented the universal action of gravitation, on the grounds that he wrote, in the last extract, of not being committed to any causal hypothesis, and later of gravity acting 'as if' it acted according to Hall's hypothesis. Though I would hold this position, it should be pointed out that Newcomb still committed himself to a certain extent. He could have calculated the motions 'as if' there was a suitable distribution of matter surrounding the Sun yet not committing himself to its actual existence. This would have had the effect of enabling him to escape the improbable value of the solar parallax and the 'ring', as a purely abstract entity, could not be attacked as being physically improbable. However, matter distributions were part of Newcomb's common vocabulary—a hypothetical abstract ring would have been absurd to him, whereas a physical law was already an abstract concept.

The change in the distance dependence from an inverse power of 2 to one of 2.000 000 1574 was a very minor change, unless one holds that only integral powers are acceptable or that the inverse square power is fundamental, and as we shall see in Chapter 5 the *a priori* arguments for this power law were rather weak. There was for instance no additional variable brought in, such as velocity which entered the laws to be discussed in Chapter 6, so this was hardly a major modification of foundations. Finally Newcomb was constructing tables and so had purely pragmatic motives. His work of 1895 was a treatise of astronomy, not of fundamental physics. Hall's hypothesis will be treated from the latter point of view in Chapter 5, but Newcomb's use of it was not, I think, meant to be of theoretical importance. One of Newcomb's key words at this stage was 'simplicity', which is found in the following passage where his attitude was being expounded:

> Should we aim simply at getting the best agreement with observations by corrections more or less empirical to the theory? It seems to me very clear that this question should be answered in the negative. No conclusions could be drawn from future comparisons of such tables with observations, except after reducing the tabular results to some consistent theory, . . . Our tables must be founded on some perfectly consistent theory, as simple as possible, the elements of which shall be so chosen as best to represent the observations . . . it did not seem to me best that we should wholly reject the possibility of some abnormal action or some defect between the assumed relations of the various quantities. (Newcomb 1895, pp. 173–4)

Now adoption of the matter hypothesis nowhere involved one in inconsistency or complexity. One just used the same anomalies as Newcomb used and ascribed them to perturbing matter in the solar system. This was simple and consistent. Note though the last sentence—again defect in theory is mentioned.

It has been shown that the attempt to explain the motion of the nodes of Venus by decreasing the mass of the Earth was improbable. Now this type of reasoning was quite usual in celestial mechanics when comparing observation with theory. We have already seen how in 1882 Newcomb used the improbability of a large mass of Venus to reject the possibility that the advance in the perihelion of Mercury was due to too small a mass of Venus. Now Newcomb was faced with two choices. He could try and explain the advance in some other way or he could say that it must be due to error—a choice that he mentioned many times. Now as regards the first choice any explanation using extra perturbing matter was going to introduce discrepancies elsewhere and no hypothesis change could have given a motion to the nodes as it could to the perihelion and aphelion. This first choice then forced one back to the full matter hypothesis and a retraction of Hall's hypothesis. The second choice was not so improbable. The most famous example of this in the nineteenth century was the discovery by John Couch Adams in 1853 that Laplace was wrong when he thought that the secular acceleration of the Moon's mean motion could be completely explained by theory. It had been one of Laplace's great triumphs when he showed that the 12″ acceleration could be explained by the decreasing (long period) eccentricity of the Earth's orbit, but Adams (1853) showed that the theoretical value had to be decreased by 1″.66, and by further orders of approximation he later showed (Adams 1859) that theory in fact gave only 5″.64, only half the observed value. The discrepancy soon came to be regarded as due to tidal friction effects. Now Newcomb mentioned the possibility of error a number of times, in fact so many that he seemed often on the verge of accepting it. However, if he had accepted it, it would have involved him in difficulties with the matter hypothesis, and it is possible that it was for this reason that he did not make this choice. We saw that the strongest argument that Newcomb had against the matter rings was the improbability of their having the required high inclination. Now this high inclination was required mainly to explain the advance in the nodes of Venus. If Newcomb decided that this advance was due to some mistake in constructing the theory or combining the observations the requirement for the high inclination was removed and with it his only strong argument against the matter hypothesis.

The first choice then led to the matter hypothesis, and the second led either to Hall's hypothesis or the matter hypothesis (ring of small inclination). Since the latter was not improbable, we must consider the prior probability of Hall's hypothesis before making any decision between the two.

If one looks at the theories that were being considered at the end of the nineteenth century as tentative explanations of the inverse square law of attraction, and this will be done in detail in Chapter 5, one sees that all gave

Clairaut hypotheses. The additional term or terms were all very small and close approximation to the inverse square law was obtained, but where arbitrary coefficients were involved the possible alterations led to a Clairaut law and not to that of Hall. On these grounds then Hall's hypothesis was *a priori* improbable. As far as agreement with the inverse square law was concerned—and its successor had to agree with it where it was well confirmed—Newcomb devoted a short section to showing that it had the same attraction over distances from less than a metre to the distance between the Sun and Uranus. Further than this, however, 'the result of observations so far made is relatively less precise' (Newcomb 1895, p. 120). This applied to Neptune and to the distances between the components of physically connected double stars. He did mention an effect which could have been used as being favourable to the probability of Hall's hypothesis but which was not here taken as such:

> Although the total action of a material point on a spherical surface surrounding it would converge to zero when the radius became infinite, instead of remaining constant, as in the case of the inverse square, yet the diminution in the action upon a surface no larger than would suffice to include the visible universe would be very small. (Newcomb 1895, p. 120)

However, Seeliger (1895) did not think that Hall's hypothesis was in any better position than the inverse square law as he considered the potential which goes to infinity in both cases; because Seeliger is of central importance in the development of the zodiacal light hypothesis the argument will be considered more closely in the next chapter.

Accepting that Seeliger's assertion that Hall's hypothesis suffered from this cosmological objection as much as the inverse square law and having seen that Hall's hypothesis was theoretically improbable (and this does not change if one remembers that the theories purporting to explain gravitation were themselves improbable), one is led to the conclusion that it was *a priori* improbable. Accepting Newcomb's version of the cosmological argument and taking the attitude that Hall's hypothesis is such a minor alteration to the inverse square law that it cannot be considered as improbable, one must allow this hypothesis a certain degree of probability. The position that Newcomb found himself in was thus quite a difficult one. He accepted the validity of the advance of the nodes of Venus and used Hall's hypothesis; this involved the improbability of the solar parallax value. If he had taken a matter ring he would have invoked the improbability of a high inclination. The alternative was to take the advance of the nodes of Venus as being due to error and perhaps to insert an *ad hoc* factor to account for it temporarily. In this alternative the matter ring can be of small inclination, so that it is not improbable on that score. However, one can interpret Hall's hypothesis as not improbable as well, so that taking the motion of the nodes of Venus as due to error leads to hypotheses which appear probable whereas accepting it as valid leads to improbable hypotheses. This might have been grounds for declaring it to be due to error; (the resolution of this problem will be recounted in the next chapter), but it can be said now that it would have supported this declaration.

Newcomb did not have to wait long before new evidence appeared in the form of Brown's new lunar theory. The complete theory was published in 1903, but by 1897 he had announced the preliminary results for the motions of the lunar perigee and nodes (Brown 1897). These were found to be $1''.0 \pm 1''.8$ and $0''.2 \pm 1''.1$ per year respectively. The large errors forced these results to be accepted equivocally for it was possible to fit the perigee motion of Hall's hypothesis, $1''.4$ per year, as well as no perigee motion at all. However, by 1903 Brown's results were firmly against Hall's hypothesis (Brown 1903). The annual motions of the node and perigee were found to be $0''.1 \pm 0''.2$ and $-0''.1 \pm 0''.2$ respectively, which left no room for the inverse $2+\delta$ power law:

It appears, then, that this assumption must be abandoned for the present or replaced by some other law of variation which will not violate the conditions existing at the distance of the Moon. (Brown 1903, p. 397)

Newcomb recognized the force of Brown's results, and in a short discussion at the end of the last paper written before his death in 1910 he revised his opinion:

After the publication of my tables, Brown's completed determination of the theoretical motion of the lunar perigee showed that the gravitation of the Earth did not differ from the Newtonian law and the same thing was true presumptively of that of the Sun. We were thus forced back upon the hypothesis of a mass of matter surrounding the Sun sufficient to cause the motion of the perihelion of Mercury. (Newcomb 1912)

He did not revert to the hypothesis of a matter ring that he had considered in 1895 but thought that Seeliger's hypothesis of matter ellipsoids was preferable. This will be considered in the next chapter, but it is worth noting that in 1882 Newcomb had suggested the argument that was later used to overthrow Seeliger's hypothesis when he queried whether a mass sufficient to perturb Mercury to the extent required would be as invisible as it appeared to be. At the time he thought that this could not be quantified, and it was not done so until early in 1915.

Finally Newcomb referred to the nodes of Venus.

All the determinations of the elements were made on the hypothesis that the motion of the perihelion of Mercury was not caused by the attraction of matter. Introduce this matter as an attracting body and we have a new set of secular variations, especially of the node of Venus. This may well be why the value of the solar parallax which I derived from the node of Venus, notwithstanding its seeming certainty, is abnormally small. (Newcomb 1912, p. 227)

This is what was argued above.

4

SEELIGER'S HYPOTHESIS OF INTRA-MERCURIAL MATTER

4.1. An alternative intra-Mercurial matter hypothesis

When Simon Newcomb was forced to abandon Hall's hypothesis as an explanation for the anomalous advance in the perihelion of Mercury, owing to the failure of the hypothesis to give the correct lunar motion, he had to return to the Newtonian inverse square law and a material explanation of the advance. By this time a number of scientists had suggested radical gravitational theories, but none had really provided an alternative to the Newtonian scheme which was coherent and convincing and at the same time capable of giving the full perihelion advance. In 1859 Le Verrier had proposed that the advance was caused by a large number of asteroids surrounding the Sun. This idea had eventually failed through a lack of observational support, although the attempts to corroborate it centred round the possible existence of the planet Vulcan. The new matter hypothesis to which Newcomb transferred his support was proposed in the opposite way to the earlier one. Now the visible evidence was presented first and fitted in to a scheme of matter rings which, it was shown, could just about account for all the anomalies in the planetary motions that Newcomb had found. The visible evidence was known as the zodiacal light. This was a very faint cone shaped area of light, visible at dusk and dawn in favourable conditions, rising from the point where the Sun had set or was about to rise and pointing along the zodiac. The faintness of the light made observation difficult and we will see that data concerning it was often conflicting. Astronomers now generally accept that the light is caused by sunlight which has been reflected or scattered off a large number of minute particles surrounding the Sun. In the middle of the nineteenth century a number of contrasting explanations had been advanced, and no one had suggested the zodiacal light as a candidate for Le Verrier's intra-Mercurial matter.

The astronomer who convinced most scientists that the zodiacal light was responsible for the anomalous advance in the perihelion of Mercury was the German Hugo von Seeliger. He was born in 1849 and died in 1924. From 1882 until his death he was professor of astronomy and director of the Munich Observatory. Seeliger was one of Germany's most influential astronomers, both as a supervisor to a large number of astronomers taking their doctorates and as president of the Astronomische Gesellschaft from 1896 to 1921. It was in 1906 that he published his zodiacal light hypothesis. In the absence of any convincing alternative and with few astronomers prepared to explore tentative ideas of the radical physicists, the astronomical establishment soon accepted Seeliger's hypothesis. It was a successful theory as it was widely accepted, explained all that it was required to, and was observationally verified. As support for general relativity became established, from late 1915 onwards, Seeliger's hypothesis faded from view, although inevitably some scientists never relinquished Newtonian theory and retained their belief in the zodiacal light explanation of Mercury's perihelion advance. At the end of this chapter we will consider three prominent astronomers and the circumstances in which they eventually rejected Seeliger's hypothesis—Freundlich (1913), de Sitter (1916), and Jeffreys (1919).

The alternatives to Seeliger's hypothesis that were available at the turn of the century will be considered in Chapters 6 and 7. They came from a number of new gravitational theories, and their advocates were mainly physicists. For the practising astronomer the zodiacal light hypothesis was all that was required. Before discussing it in detail, we need to outline preceding ideas on the zodiacal light and the early work of Hugo von Seeliger.

4.2. The zodiacal light

The Italian astonomer, J.D. Cassini, who moved to Paris in 1669 and controlled the new Paris Observatory, is generally recognized as being the first to have attached significance to the zodiacal light, to have made systematic observations of it, and to have put forward an explanatory hypothesis for it. Before Cassini a few scattered references may be found. Jones (1856) gave a short list from Pliny to the English naturalist Childrey, who recommended the light 'to the observation of mathematical men' (Childrey 1661). Childrey noted that it was visible in northern latitudes in February and March just after sunset, but he did not know that it is also visible in these latitudes before sunrise in the autumn. Cassini first noticed the zodiacal light in 1683 and almost immediately proposed a hypothesis to explain the phenomenon (Cassini 1730). He suggested the existence of a large number of small particles surrounding the Sun in the region of the ecliptic with sunlight being reflected off these particles or their being self-luminous. The matter had been ejected from the Sun and was possibly of the same nature as the matter forming the sunspots and the faculae. The plane of the light appeared to lie in the plane of the Sun's equator. The connection with the sunspots gave a possible clue to the irregularities in the visibility of the zodiacal light, the spots being well known for their irregularity of appearance. Cassini noted that in 1688 the light started to diminish in intensity and that since that date no more sunspots had been observed.

Cassini introduced the analogy of the Milky Way shining like a diffuse cloud and yet composed of many stars—the zodiacal light was a diffuse light due to the combined effect of many shining '*petites planètes*' (Cassini 1730, p. 206). The use of the word 'planet' in the analogy indicates that the zodiacal light particles not only were ejected from the Sun but then moved round it like Mercury and Venus. Cassini, who is usually represented as having been too much in fear of the Church to have publicly accepted the motion of the Earth (Dreyer 1953, White 1960), added that each of the Ptolemaic, Copernican, and Tychonic systems could represent the movements of these little planets. The motion of the zodiacal light matter round the Sun was possibly added by Cassini under the influence of a letter to him written by Nicolas Fatio de Duillier (Cassini 1730, p. 157). Fatio invoked large and irregular particles moving round the Sun and reflecting its light. The Fatio–Cassini hypothesis is similar to what is held today, though the sunspot connection has been discarded and gravitation is brought in to control the circumsolar movement. A standard pre-war textbook of astronomy takes as sufficient 'particles one millimetre in diameter, of the low albedo of the moon, and at an average distance of about five miles apart'.(Russell, Dugan, and Stewart 1928, p. 359)

Other hypotheses were at times proposed. In 1731 Mairan suggested that the light was due to the Sun's atmosphere. Later Laplace showed that an atmosphere fixed relative to the central body could not extend beyond a certain distance where the centrifugal force overcame the gravitational attraction. In the 8th edition of the *Encyclopaedia Britannica* Robert Main (1853) wrote that before this result Mairan's hypothesis was the one 'generally adopted'. Interest in the zodiacal light was never very great, but it increased in the nineteenth century when Biot suggested that the prominent meteor shower of 1833 was caused by the Earth's passing through the node of the matter causing the light. In 1844 Houzeau attempted to show that this was not so and intimated that the cause of the light might be 'more local than hitherto supposed'. This suggestion was developed by George Jones who had made extensive observations of the zodiacal light while sailing in the Pacific as an American Navy Chaplain and who proposed that the light was due not to a circumsolar ring of matter but to a circumterrestial ring of matter (Jones 1855, 1856). This hypothesis was attacked by Barnard (1857) who pointed out that the axis of the light underwent no parallactic displacement no matter how far the observer moved and that Jones's involved explanation of this was unsatisfactory. This and other criticisms would seem to dispose effectively of the terrestial ring hypothesis, though Pringsheim (1910) in his book on solar physics wrote that 'many investigators of course ascribe the zodiacal light to a ring suspended round the Earth'.

Matter hypotheses were not the only type to be proposed. Balfour Stewart suggested that the light was due to electrical discharges in the Earth's atmosphere (see Proctor 1870, 1872), and Brewster (1845) suggested that the passage of the Earth left a visible trace in the ether. Challis (1863) attempted to fit the phenomenon into his hydrodynamical scheme by ascribing it to the interaction of the Sun's proper motion and the Sun's rotation about its own axis and the disturbing effect of this interaction on the ether.

The zodiacal light itself is difficult to observe, and if one compares Barnard's (1857) attack on Jones with that of Piazzi Smyth (1857), one might think that different phenomena were being discussed. For instance Jones had used as supporting evidence for his hypothesis the fact that he had seen the light in both the east and the west at the same time when the Sun was more than 90° below the horizon. Piazzi Smyth attacked these observations and said that he had never seen such a phenomenon. Yet Barnard accepted them, and indeed they become the grounds for asserting that the zodiacal light matter extended beyond the orbit of the Earth. Furthermore, in 1854 Brorsen had announced the discovery of the *Gegenschein*, a faint patch of light directly opposite the Sun, and it became apparent that there was a zodiacal arch, i.e. a faint band of light running through the zodiac increasing in brightness in the middle thus forming the *Gegenschein*. This did not support Jones's theory for according to that the *Gegenschein* ought to be brighter than the light at the horizon. Even the inclination of the axis of the light was contentious. Some said it lay along the ecliptic—this was 'well known' for Proctor (1872, p. 164) and accepted by Newcomb (1895, p. 115). However, Cassini (1730, p. 143) thought that it was inclined along the plane of the Sun's equator, which is about 7° to the ecliptic, and Seeliger also held this following two sets of observations (Wolf 1900, Marchand 1895). Finally Wright (1874) showed that the zodiacal light was polarized in a plane passing through the Sun and that its spectrum was the same as that of the Sun, and concluded that

'the light is derived from the Sun, and is reflected from the solid matter. This solid matter consists of small bodies (meteoroids) revolving about the Sun in orbits crowded together toward the ecliptic.' (Wright 1874, p. 458)

This then was the situation regarding the zodiacal light in which Seeliger found himself at the turn of the century. Perhaps the crucial difference between him and Newcomb was his acceptance of a 7° inclination for the zodiacal light matter. Newcomb had used as one of his arguments for rejecting matter hypotheses the high inclination required for the matter on grounds of dynamical improbability. Since he had direct observational evidence for this inclination, Seeliger did not need to concern himself with this argument.

The discussions of the years 1906 to 1919 ranged around the question of what mass might be assigned to the zodiacal light matter, the existence of the matter itself being taken as proven. The question was usually discussed in an alternative form, i.e. whether the matter, having the mass required to explain the Newcomb anomalies, could have the low luminosity of the zodiacal light. We thus find a return to the questions of observation which in a different form had so plagued Le Verrier.

4.3. Seeliger and the inverse square law

Though Seeliger chose a matter hypothesis in 1906 to explain the anomalous advance in the perihelion of mercury, he had earlier tried the alternative approach to the problem, i.e. changing the inverse square law. However, this was done for quite another problem, the cosmological objection to the inverse square law, an objection which Seeliger was responsible for uncovering. He first published this in 1895 (Seeliger 1895) and then in more detailed form in 1896 (Seeliger 1896). The latter paper cited Neumann's *Allgemeine Untersuchungen über das Newton'sche Princip* (Neumann 1896), in which Neumann pointed out in reference to Seeliger's 1895 paper that he himself had indicated the problem much earlier (Neumann 1874, p. 502). This though was a passing reference in a paper concerned with electrodynamics. Seeliger's problem was the following. For a potential function $f(r)$ and a density distribution of mean value δ, the total potential at a particular point is given, using spherical co-ordinates, by

$$V = \int_0^{2\pi} \mathrm{d}\phi \int_0^{\pi} \sin\gamma \, \mathrm{d}\gamma \int_{R_0}^{R_1} \delta f(r)\, r^2 \, \mathrm{d}r$$

Now for the Newtonian case we have $f(r) = 1/r$, so that the integrand is $\delta\, r$. Taking δ as finite and passing to the case of an infinite universe, i.e. taking R_1 to infinity, one finds that the value of V becomes infinite so that the value of the gravitational force becomes indeterminate†. The situation is the same for Hall's hypothesis, or, as Seeliger called it, Green's formula (following George Green's earlier discussion of the formula), since then the integrand is $\delta \cdot r^{1-n}$, $n \ll 1$, which still becomes infinite. Neumann considered a formula of the form

$$f(r) = A \exp(-\lambda_1 r) r^{-1} + B \exp(-\lambda_2 r) r^{-1}$$

which was not subject to this objection. Nor was the formula which Seeliger seemed to favour, this being the product of the inverse square law and a factor $\exp(-\lambda r)$. This law, said in 1896 to be 'only an example', nevertheless was introduced in 1894 by a serious analogy with the problem of Cheseaux and Olbers concerning the night sky (Olbers' paradox). Olbers had suggested as solution to the problem (why, in an infinite universe, is the night sky dark?) that light was absorbed in its passage through space, so that rather than the intensity decreasing as the inverse square of the distance it decreased as the product of the inverse square and a factor $\exp(-\lambda r)$.

Now a change in the inverse square law had obvious possibilities for explaining the advance in the perihelion of Mercury, and Seeliger did not fail to take advantage of this. He compared the perihelion advances found by Newcomb with those expected from Neumann's law, Green's law, and the

† One might argue that since the force has a well-defined limit it does not matter that the potential goes to infinity.

TABLE 4.1

Planet	Neumann	Green	Laplace	Newcomb
Mercury	40	40	40	40
Venus	55	16	29	-8 ± 37
Earth	64	10	25	6 ± 8
Mars	79	5	20	8 ± 4
(Moon)	—	156	0.9	—

The units are all seconds of arc per century. From Seeliger 1896, p. 386.

exponential law (Laplace's law), the adjustable parameters being evaluated to fit the advance of Mercury. The values are shown in Table 4.1.

Neumann's law was hopeless. Laplace's law was discrepant for the outer planets and its favourable value for the lunar perigee could not be taken into account before Brown's theory of 1903. For various reasons Green's law was equivocal on the lunar perigee (Newcomb 1895, p. 119). However, Seeliger refused to adopt Green's law for two reasons. As we have seen it suffered from the cosmological objection—it did not fit an infinite universe. Secondly it failed to account for the motion of the nodes of Venus, the second in size of Newcomb's anomalies, and it was implied that any explanation of that would cover the perihelion advances as well. This left the way open for the later matter hypothesis, and it also did not preclude Seeliger from adopting Laplace's law if he so wished: the planetary application used $\lambda = 0{\cdot}000\ 000\ 38$, and since $d\varpi \propto \lambda$ one could take say $\lambda = 0.000\ 000\ 01$, satisfy the planetary motions, and still not be subject to the cosmological objections.

Though it appears that this was what Seeliger finally did (see below) he never committed himself. He was above all bound to the view of an infinite universe, but this cosmological problem and the related problem of Cheseaux and Olbers can be evaded in a number of different ways. Since this was one of the reasons that Seeliger rejected Green's law/Hall's hypothesis, after Newcomb's acceptance of it and before Brown's lunar refutation of it, it seems appropriate that we should consider this in a little more detail.

4.4. The cosmological objection

Seeliger's 1896 paper ended with the claim that the darkness of the night sky did not necessarily involve invoking Olbers' explanation of light absorption. For this he has recently been attacked by Jaki who, having seen the importance of Olbers' paradox in modern cosmology (if it is that important—see Harrison 1974), thinks that everyone before should have realized its importance and thus adopted the idea of a finite universe (Jaki 1969). Seeliger though had little reason to be anything other than conservative in his approach, an approach which was outlined in a paper published in 1909 entitled *On the application of the laws of nature to the universe* (Seeliger 1909). In this he considered the applicability of the Newtonian law to the universe as a whole, the problem of Cheseaux and Olbers, and the applicability of Clausius's idea that the entropy of the universe tends towards a maximum. The Newtonian gravitational case led to the following pair of statements.

If the law of Newton is absolutely exact, the infinitely extended parts of space cannot be filled with matter of which the mean density is finite.

I have thus expressed the converse of the theorem:

If the total mass of the universe (the mean density being finite) is infinitely large, Newton's law cannot be considered as the exact mathematical expression of the attractive forces that are acting. (Seeliger 1909, p. 93)

Note that the possibility of a finite universe was not even considered. The problem here was the difficulty of conceiving of a bounded universe, but with the notion of curved space one could have a universe that was finite and unbounded. Karl Schwarzschild had shown in 1900 that this was not inconsistent with measurements of stellar distances but Seeliger did not refer to it.

The first of Seeliger's conditions involved two choices. The first was that there was only finite matter in the universe, but this was problematic when an infinite universe was assumed. On the other hand it was possible to have a finite mean density without a finite amount of matter if the infinite amount was arranged in certain ways. Seeliger wrote that this was easy to do, and gave an example. His use of 'easy' echoed that of John Herschel in a letter to Richard Proctor concerning the problem of the dark night sky (Proctor 1891). Herschel did not like Olbers' hypothesis of the extinction of light in its passage through space (Herschel 1851) and suggested how a suitable system could be built up. The letter was prompted by a similar suggestion of Proctor published in 1869. In *Other worlds than ours* Proctor again discussed a hierarchical system consisting of

first satellite-systems, then planetary-systems, then star-systems, then systems of star-systems, then systems of systems of star-systems, and so on to infinity. (Proctor 1891, p. 274)

All that one was subject to was a condition as to the ratio of the number of shells in each system to the brightness loss from one system to the next, and the bright night sky might be evaded. Proctor used this to argue that the nebulae were inside our galaxy and that the members of the next higher system were not visible to us, these views being supported by other arguments. A hierarchical universe was proposed as the solution to Olbers' paradox and to Seeliger's gravitational objection by Charlier in 1908 (and repeated in 1922) who was supported by Selety (North 1965).

Charlier's hierarchical system was attacked by Arrhenius for being unable to account for the formation of single galaxies and for not having any mechanism of stability. Seeliger also objected to hierarchical universes, not on physical grounds but from a wish to be as independent as possible of metaphysical speculations. The metaphysical idea involved in the acceptance of the exact nature of the inverse square law and a mean finite density of matter in the universe was the necessity of inferring the arrangement of matter in distant parts of the universe, an arrangement that we could not see from the behaviour of matter locally:

> To declare that the law of Newton is an absolutely exact law of nature means supposing that condition (1) is satisfied, which will imply that, from experimental states obtained in a narrow domain one pronounces on properties of matter in regions of the universe infinitely far away. The constitution of matter in our neighbourhood must therefore be determined *essentially* by the distribution of matter at an infinite distance. It is scarcely useful to insist on the absurdity of these consequences based ultimately on hypotheses which escape completely and for ever our knowledge. (Seeliger 1909, p. 95)

Now the elimination of metaphysics is very Machian, and it is strange that the 'absurd' second sentence bears a strong resemblance to Mach's Principle. Of course it was the 'infinite distance' that was metaphysical, but Seeliger did not eliminate it—he wished to retain an infinite universe—but only eliminated the need to refer a physical property (arrangement of matter) to such a metaphysical construct.

It is surprising that Seeliger did not refer to a suggestion by Föppl (1897) that would have enabled the whole problem to be bypassed. One cannot claim that Seeliger had not known of the paper for it was he who read it out to the Akademie der Wissenschaft in Munich in 1897. Föppl suggested that there existed negative masses as well as the usual 'positive' ones. This not only provided a solution to Seeliger's problem, it also brought the theory of gravitation into line with the theory of electricity with its positive and negative charges and furthermore was quite plausible on the hydrodynamical theories of gravitation then being considered. The absence of negative masses in our neighbourhood could be explained by the repulsion between them and our own system; negative systems would look the same as positive ones, i.e. the two stars in a physically connected binary each of negative mass would move round each other in the same manner as if they were positive masses. Föppl's suggestion seems to have had little impact, and the only person who linked it to other astronomical phenomena was Schuster (1898*a*). He wondered how the rotation had been imparted to the solar system. The conservation of angular momentum would be obeyed if another system had been given an equal and opposite amount, but the problem then arose as to what happened to that system. If its mass was of the opposite kind to ours, it would have been repelled and we would not expect to see it near to us. Furthermore, negative masses could perhaps explain the large proper motions of some stars (e.g.

Groombridge 1830), the velocities of which Schuster took Newcomb as holding to be too great to be explained by gravitational attraction alone. These thoughts were in fact published by Schuster as a *Holiday dream* in the August issue of *Nature* in 1898, but in the October issue he referred them to the more serious base of Pearson's sources and sinks and to Föppl's positive and negative masses (Schuster 1898*b*). Schuster's queries themselves are not of great weight. It is usually considered unproblematic that two stars may nearly collide, producing the beginnings of a planetary system, angular momentum being conserved, and then continue on their way at appropriate velocities. This is the basis of Chamberlin and Moulton's theory of the origin of the solar system.

Seeliger, having now been forced to accept a change in the inverse square law, concluded the section by discussing the change on independent grounds. Its verification was confined to the solar system (this was of course three years after Seeliger had himself explained the Newcomb anomalies in 1906), for though it appeared to apply to double-star systems the precision was less than wholly convincing. Its applicability was not known for distances of the order of those between ordinary stars. Furthermore, according to those theories in trying to explain the action of gravitation the inverse square law only approximated to the law of attraction. In particular Seeliger cited Le Sage's theory, revived by William Thomson, and mathematically discussed by George Darwin (1905). Darwin found the actual law of attraction on this theory to be a complicated Clairaut-type law (i.e. a function of second and higher inverse powers of the distance) which did not help the cosmological objection. However, this theory gave the possibility of gravitational absorption which did help this problem. Now laboratory experiments had been carried out to test for absorption but, not surprisingly, had failed to detect any effect. Later Majorana (1920) claimed to have detected such an effect in a laboratory—he surrounded a leaden ball, of weight 1274 g *in vacuo*, by 104 kg of mercury and found that the ball had lost $0.000\,98 \pm 0.000\,16$ mg in weight! Silberstein (1924), with restraint, called these 'untimely claims'. It was felt that a much more likely test would be afforded by lunar eclipses. In fact there were discrepancies in the Moon's motion—periodic fluctuations in longitude discovered by Newcomb. A pupil of Seeliger, K.F. Bottlinger, investigated the question and in 1912 announced that the Newcomb fluctuations were well represented by the hypothesis of absorption of gravitation in the Earth (Bottlinger 1912). de Sitter, who had looked at the question in 1909, was prompted by Bottlinger's result to recommence his investigation and in 1913 showed that Bottlinger's acceptance of absorption was not well founded (de Sitter 1913*a*). In 1914 Bottlinger used Brown's lunar theory to show that absorption was not warranted and was also not observed in the motion of the Martian satellite Phobos (Bottlinger 1914). This did not mean that absorption was not effective against the cosmological objection where the important absorbers would be stars, bodies much more massive than the Earth. Seeliger had shown that the absorption coefficient had to be less than 10^{-8} in order to agree with the perihelion motions (i.e. give none), but Bottlinger's 1912 value was 10^{-15} and Majorana was working to 10^{-12}. On the Bottlinger and de Sitter results we can then say that the coefficient must be less than 10^{-15} but as we go to infinity the term $e^{-\lambda r}$ could still go to zero and the objection be overcome. The cosmological absorption hypothesis can thus only fall in hand with the LeSage–Thomson theory, which in turn falls with the Newtonian theory it was designed to support. Seeliger's programme, of course, was to obtain as much as possible from the Newtonian theory. He also used an absorption hypothesis to explain the darkness of the night sky—absorption that was in the ether as well as owing to the 'interposition of dark or less brilliant bodies', and 'extended clouds of cosmic dust' (Seeliger 1909, p. 98).

Seeliger also mentioned the importance of the finite life of stars and the finite velocity of propagation of light. William Thomson used these factors to give a solution to Olbers' paradox. In his *Baltimore Lectures* Kelvin, as he had by then been ennobled, computed the ratio of the apparent brightness of the night sky to the brightness of the Sun's disc (Kelvin 1904). By taking the standard observation that there were 1000 million stars within a parallax of $10^{-3''}$ (3.09×10^{16} km), this quantity α was found to be 3.8×10^{-13}. Now Cheseaux and Olbers felt that this quantity ought to be equal to unity; α itself is proportional to r so that this radius of the visible stellar sphere can easily be calculated. Kelvin chose $\alpha = 0.0389$, or about 4 per cent, for which r is 3.09×10^{27} km. Now light requires 3.27×10^{14} years to travel this distance, and it was this that enabled Kelvin to turn aside the impact of Olbers' paradox for it is much longer than the apparent lifetime of a star. However, his lifetime estimates were based on stellar energy being due to gravitational contraction. Though this was the only available energy source that gave lifetimes that were in any way acceptable, the age Kelvin assigned to the Sun, 20 million years, was in direct conflict with geological evidence which indicated that the Earth was 250 million years old and which was needed by the theory of natural selection. When radioactivity was discovered it was felt that this might enable Kelvin's estimate to be substantially increased (de Tunzelmann 1910). There was some indirect evidence for this as helium was observed in the Sun and was a decay product of radium. The insufficiency of radioactivity as the source of the solar energy was pointed out by Lindemann (1915). Nevertheless the long time scale (Jeans in his Adams Prize Essay of 1917 (Jeans 1919) suggested 560 million years as the time since the primeval nebulae broke up into stars which would agree with the geological time if the solar system did not finally form for another 200 million years) could then be covered by a suitable energy source. The mass–energy relation of special relativity enabled extremely long stellar lifetimes to be envisaged. Jeans in his 1917 essay calculated that a 1 per cent reduction in the Sun's mass would produce enough energy to last 150 000 million years at the present rate of radiation. Jeans himself is sometimes credited as having first suggested the possibility of the creation of energy by the annihilation of matter (Waterfield 1938)† ('the coalescence of a positive and negative ion') before the results of special relativity. This was published in 1904 as a tentative explanation of radioactivity (Jeans 1904). However, the conceptual basis was the idiosyncratic theory of a granular ether proposed by Osborne Reynolds and the annihilation was the 'combination and mutual annihilation of two ether strains of opposite kinds'. Though we find Lindemann writing in 1915 that 'the origin of the Sun's heat cannot be referred to any known cause (Lindemann 1915, p. 372), the mass annihilation source of energy came to be used more and more whenever problems of stellar energy arose up until the late 1930s when Bethe and Weizsäcker proposed the conversion of hydrogen into helium by fusion as the actual source.

From the 20–50 million years of Kelvin we progress to say 100 million years using radioactivity. However, after 1905 the stellar lifetime can become much longer, and the geological time becomes small. Conversion of all of the solar matter into energy gives a future lifetime, at the present rate of radiation, of 15 000 000 million years (1.5×10^{13} years). Even this time is less than the 3.27×10^{14} years Kelvin had calculated as the time required to fulfil the 4 per cent brightness of the night sky. It does appear then that Kelvin's argument was on fairly safe ground. Now the gravitational equivalent of Olbers' paradox is more difficult for a dark star which does not contribute optically still contributes gravitationally. Furthermore, Kelvin's argument cannot be used in the strict Newtonian scheme for that assumes an infinite velocity of propagation. On the other hand, finite velocities of propagation of gravitational action were being seriously discussed and found a natural place in those laws containing velocity-dependent terms, of which Weber's law is the most familiar. These gave perihelion advances, although most required a velocity of propagation lower than the velocity of light in order to give the complete advance for Mercury. Only the law of Gerber gave the complete perihelion advance for Mercury using the same velocity as that of light, but most physicists found the law unacceptable. A discussion of these laws will be given in Chapter 6.

4.5. Seeliger's zodiacal light hypothesis

Seeliger had given two requirements in 1896 which any new gravitational law had to fulfil: it had to be not subject to the cosmological objection and to allow some explanation of the motion of the nodes of Venus which could apparently not be explained on its own. Hall's hypothesis failed to fulfil either requirement and the Weber-type laws, while possibly fulfilling the first

† Chapter 13 of this book gives a rare account of the problem of stellar energy from the time of Helmholtz up to 1938. Part of Jeans's paper can also be found in Shapley (1960).

requirement, failed to fulfil the second. Seeliger's Newtonian approach was able to fulfil both. The use of a Le Sage–Thomson theory to support the inverse square law fulfilled the first requirement through the mechanism of absorption. The anomalous motion of the nodes of Venus was then put with the other Newcomb anomalies and explained by Seeliger's zodiacal light hypothesis. It is this hypothesis that we shall now consider.

The zodiacal light was commonly held, as we have seen, to be caused by sunlight reflected from small particles finely distributed around the Sun and stretching as far as the Earth's orbit with a density decreasing with distance from the Sun. This matter obviously had some effect on the motions of the inner planets, but the quantitative effect was unknown. Seeliger's problem was to find out if this was sufficient to account for the Newcomb anomalies. As in the case of a planet without satellites, the mass of the matter could only be determined by its perturbative effect on other planets. Unlike other planets there were no *a priori* estimates of its density. There was, however, a constraint on the mass that was independent of the perturbative effects, and this was the brightness of the light. The variable parameter was the size of the particles constituting the zodiacal light matter. Whereas the constraint for Le Verrier had been that his matter was not visible, the constraint for Seeliger was that his matter was visible. At first this might not seem to be a constraint but rather striking evidence in favour of the hypothesis, but it is clear that if too great a mass is assigned to the matter it would be far brighter than we know to be the case. As far as perturbing effects are concerned the matter must not introduce anomalies where none have been observed. Not until 1913 were the effects on the Moon's motion and on the Earth's motion computed. Newcomb himself had rejected it because matter lying in the plane of the ecliptic would cause an unobserved regression in the nodes of Venus and Mercury. However, and this is an indication of the difficulty of zodiacal light observations, both Wolf and Marchand announced in 1900 that the light lay in the plane of the Sun's equator—an inclination of about 7° to the ecliptic with the ascending node at 74° (Wolf) or 70° (Marchand)—an inclination that had had prior adherents including Cassini himself. If Mercury's orbit lay in the Sun's equatorial plane this inclination of the matter would not affect Mercury's node and would advance that of Venus, exactly as required. Newcomb had also argued that such a high inclination was dynamically improbable, but this has no answer to a direct observation beyond denying the validity of that observation.

Seeliger's 1906 paper announcing his hypothesis started by giving brief discussions of the possibility of the non-sphericity of the Sun and of Hall's hypothesis, and then gave a detailed account of the secular effects of matter ellipsoids surrounding the Sun (Seeliger 1906). He considered five such ellipsoids of semi-major axes 0.10, 0.17, 0.24, 0.60, and 1.20 (Earth, 1.00; Mercury, 0.387). The three intra-Mercurial ellipsoids were found to have similar effects, so he chose the third alone, and as the fourth was found to be redundant the final assumption was of two ellipsoids, numbers 3 and 5.

A new consideration was then introduced, one that ruined the homogeneity of Seeliger's solution and appears most improbable, which was that the planetary inertial system rotated with respect to the fixed stars. This had been suggested in connection with the Newcomb anomalies by Anding (1905) in the previous year. Any specification of position requires axes 'fixed' to the fixed stars. Allowance must be made for precession, and this is done by referring co-ordinates to a particular date or epoch, say 1860.0, and then adjusting to the required date by use of the constant of precession. Though the fixed stars are not fixed, but have proper motions of their own, the same value of the constant of precession is obtained from stars of different proper motions and of different magnitudes. The co-ordinate system that is used in applying the Newtonian laws is one in which these laws hold true, and, following Ludwig Lange in 1885, we call such systems 'inertial systems'. Usually the fixed star system is taken to be an inertial system, but Anding pointed out that if the star system was not inertial but was rotating with respect to the inertial system of our planetary system this rotation would be apparent in the motions of the planets. The perihelia and nodes would all appear to rotate in the same direction, and since this was largely what occurred in the Newcomb anomalies he tried to evaluate that rotation which removed the anomalies. Since the advance of the perihelion of Mercury was so dominant, fitting the rotation to that anomaly would introduce discrepancies elsewhere so it was left out of the scheme. The Newcomb anomalies and the Anding revisions are given in Table 4.2. Of the important anomalies, the nodes of Venus and the perihelion of Mars were both improved and become less than their respective errors. The nodes of Mercury were also reduced to below their error, and the inclination

TABLE 4.2

		Newcomb anomaly (arcsec)	Anding (arcsec)
Mercury	Δi	$+0.38 \pm 0.80$	$+0.36$
	$\sin i\Delta\Omega$	$+0.61 \pm 0.52$	-0.31
Venus	$e\Delta\varpi$	-0.05 ± 0.25	-0.10
	Δi	$+0.38 \pm 0.33$	$+0.34$
	$\sin i\Delta\Omega$	$+0.60 \pm 0.17$	$+0.15$
Earth	$e\Delta\varpi$	$+0.10 \pm 0.13$	-0.02
Mars	$e\Delta\varpi$	$+0.75 \pm 0.35$	$+0.07$
	Δi	-0.01 ± 0.20	-0.03
	$\sin i\Delta\Omega$	$+0{\cdot}03 \pm 0{\cdot}22$	-0.24

i, inclination; ϖ, longitude of perihelion; e, eccentricity; Ω, longitude of node. From Anding 1905, p. 13.

of Venus and the nodes of Mars were slightly above their respective errors. The rotation which achieved this was $7.3'' \pm 2.3''$ per century about an axis perpendicular to the ecliptic, rotations about the two axes orthogonal to this being zero. The advance of the perihelion was quantitatively the same as the rotation, so that this was obviously quite hopeless for Mercury. Anding concluded his paper on a doubtful note. How did one explain this rotation of the fixed star system? Why was the constant of precession the same for near (bright) stars and distant (faint) stars? Until new stellar-statistical determinations were made, he ended, the reasons for the anomalies must be sought in the planetary system itself.

Seeliger took this rotation as a possible partial solution to the anomalies and thus had three unknowns to hand, the value r of the rotation and the densities q_3 and q_5 of the two matter ellipsoids. The orientations of the ellipsoids were also unknown, and these contributed two further unknowns, the inclination J and the node Φ of the third ellipsoid, the eccentricities having been assumed and also assuming the same inclination and node for the fifth ellipsoid. A preliminary determination gave values for the five unknowns, in particular $J = 6^\circ.95 \pm 0^\circ.97$ and $\Phi = 40^\circ.03$. These compare fairly well with Newcomb's tentative values of $J = 9^\circ$ and $\Phi = 48^\circ$ given in his 1895 discussion (Newcomb 1895, p. 114). Seeliger's value for the inclination corresponded to that of the Sun's equator, but the agreement of Φ with the Wolf–Marchand values of about 70° is not good. By introducing functions of q_0 (Sun's density), q_3, and q_5, namely B_3 and B_5, Seeliger produced a final least squares evaluation of B_3, B_5, and r. The set of these perturbations are shown in Table 4.3 which gives the contributions of each together with the total and the comparison with Newcomb's anomalies. The unknowns corresponding to these figures are $q_3 = 2{\cdot}18 \times 10^{-11}\, q_0$, $q_5 = 0.31 \times 10^{-14}\, q_0$, and $r = 5.85'' \pm 1.22''$. His rotation

TABLE 4.3

	Planet	B_3	B_5	r	Total	Newcomb	Difference
$e\,\frac{d\varpi}{dt}$	Mercury	+7.396	−0.108	+1.203	+8.49	+8.48±0.43	−0.01 (−0.01)
	Venus	+0.015	−0.009	+0.040	+0.05	−0.05±0.25	−0.10 (−0.10)
	Earth	+0.012	−0.037	+0.098	+0.07	+0.10±0.13	+0.03 (+0.01)
	Mars	+0.014	+0.033	+0.546	+0.59	+0.75±0.35	+0.16 (+0.19)
$\sin i\,\frac{d\Omega}{dt}$	Mercury	−0.049	−0.016	+0.713	+0.65	+0.61±0.52	−0.04 (−0.01)
	Venus	+0.088	+0.144	+0.346	+0.58	+0.60±0.17	+0.02 (0)
	Mars	+0.014	+0.030	+0.189	+0.23	+0.03±0.22	−0.20 (−0.18)
$\frac{di}{dt}$	Mercury	+0.574	−0.057	—	+0.52	+0.38±0.80	−0.14 (−0.11)
	Venus	+0.159	+0.009	—	+0.17	+0.38±0.33	+0.21 (+0.18)
	Mars	+0.003	−0.020	—	−0.02	−0.01±0.20	+0.01 (0)

All values are in units of arcseconds per century. From Seeliger 1906, p. 620.

was slightly less than that of Anding, and the outer ellipsoid, the one that reached outside the orbit of Earth, was very tenuous. One can see from Table 4.3 that the inner ellipsoid was mainly responsible for the perihelion advance of Mercury and the outer ellipsoid for the nodes of Venus, and that the rotation was required for the perihelion of Mars though it affected substantially the perihelion of Mars and the nodes. In fact the outer ellipsoid could be said to be redundant, since its main effect, causing the perihelion of Mars to advance, was only a third of the contribution of the rotation. However, this ellipsoid was the one for which there was the independent optical evidence of the zodiacal light being visible on the east and west horizons at the same time so that it could not be rejected altogether.

The Newcomb anomalies were all well represented by Seeliger's hypothesis. Independent optical evidence seemed to confirm the presence of matter in the position required. The implausible part of the hypothesis was the rotation of the system of the fixed stars relative to the planetary inertial system. The value of the rotation was less than that due to Anding, and Seeliger thought that this reduction was a step in the right direction. Table 4.3 shows that if the rotation were excluded the perihelion of Mars would become discrepant, and the question arose as to whether one could increase the mass of the outer ellipsoid so as to cancel the effect of excluding the rotation.

This question was answered in the affirmative by de Sitter (1913b). He thought that the rotation was '*a priori* extremely improbable' (de Sitter 1913*b*, p. 299). He started by reducing the anomalies slightly by using a larger constant of precession than that used by Newcomb, and taking the mean of this value and that found by Boss. This change had the same effect as a rotation of $1''.24$ so that the perihelion and nodes each advanced by this quantity per century. The values of q_3 and q_5 which were now required to give the anomalies were 2.42×10^{-11} and 0.37×10^{-14} respectively. The remaining differences are given under A in Table 4.4. de Sitter also considered the effects of special relativity which he had discussed with respect to astronomy in 1910. There he had found that the only sensible change was a $7''.15$ advance in the perihelion of Mercury. Since it was the inner ellipsoid that was mainly responsible for this perihelion, its density could be slightly decreased, but only to 2.09×10^{-11}. The relativistic differences are shown under B in Table 4.4. The differences for A and B were so similar that no decision could be taken on the merits of these theories using these alone. The differences were only slightly worse than those of Seeliger. Three of them were greater than the error in the Newcomb anomaly, compared with none for Seeliger. No one attempted to explain the eccentricity anomalies, two of which were greater than their errors, except for C.L. Poor in the 1920s who will be discussed later. Seeliger remarked that one could obtain such secular changes by taking higher powers of eccentricity and inclination, and implied that these four anomalies had no need of extra explanation (Seeliger 1906, p. 613). Of the fifteen

TABLE 4.4

		Mercury	Venus	Earth	Mars
$\frac{de}{dt}$	Newcomb	-0.88 ± 0.50	$+0.21\pm0.31$	$+0.02\pm0.10$	$+0.29\pm0.27$
$e\frac{d\tilde{\omega}}{dt}$	Newcomb	$+8.48\pm0.43$	-0.05 ± 0.25	$+0.10\pm0.13$	$+0.75\pm0.35$
	Seeliger	-0.01	-0.10	$+0.03$	$+0.16$
	A	$+0.00$	-0.05	$+0.18$	$+0.52$
	B	-0.02	-0.06	$+0.17$	$+0.50$
$\sin i\frac{d\Omega}{dt}$	Newcomb	$+0.61\pm0.51$	$+0.60\pm0.17$	—	$+0.03\pm0.22$
	Seeliger	-0.04	$+0.02$	—	-0.20
	A	$+0.55$	$+0.01$	—	-0.11
	B	$+0.55$	$+0.01$	—	-0.11
$\frac{di}{dt}$	Newcomb	$+0.38\pm0.80$	$+0.38\pm0.33$	-0.22 ± 0.27	-0.01 ± 0.20
	Seeliger	-0.11	$+0.18$	—	0.00
	A	-0.12	$+0.17$	—	$+0.05$
	B	-0.01	$+0.20$	—	$+0.05$

All values are in units of arcseconds per century. From de Sitter 1913b, p. 301.

anomalies, i.e. including the eccentricities, five were greater than their errors for de Sitter and two for Seeliger. However, de Sitter did not mind this, for 'according to the Gaussian theory of errors we should, amongst 15 residuals, expect 4.7 exceeding the mean error' (de Sitter 1913*b*, p. 302). Theories A and B were then just as good empirically as that of Seeliger. de Sitter did not make a choice between them; if anything one would probably have chosen A since it omitted the Seeliger–Anding rotation and did not assume the uncertain applicability of special relativity to gravitation. However, 'Seeliger's hypothesis' can be used to describe any hypothesis utilizing the zodiacal light or similar matter—though it must be a Newtonian theory and B is excluded. de Sitter ended his paper by proposing a requirement for the acceptance of Seeliger's hypothesis: the effect of the zodiacal light matter must be computed for the Moon's motion, the Earth's motion, and the motion of Encke's comet.

This was carried out by Woltjer (1914) on de Sitter's prompting. He considered the motion of the Moon and the perturbation of the obliquity of the ecliptic. The resulting motions of the lunar perigee and node were $+2''.12$ and $+2''.50$ respectively so that no discrepancy was introduced into the lunar motion. The motion of the ecliptic was found to be $-0''.501$ per century. Newcomb had found a discrepancy of $-0''.22\pm0''.18$ so that this became, adding Woltjer's result to the theory, $+0''.28\pm0''.18$, a slight increase in the discrepancy.

de Sitter (1914) used Woltjer's paper to give another discussion of Seeliger's hypothesis. He now retracted his former opinion regarding the rotation.

Without it one obtained for the secular variation of the ecliptic the 'entirely inadmissible' result of $1''.18$ compared with the Newcomb anomaly of $0''.28 \pm 0.18$ (adjusted for Woltjer's result): 'We conclude therefore that the rotation r_1 is a vital part of the explanation' (de Sitter 1914, p. 35). The only escape from this would be to decrease the inclination of the outer matter ellipsoid to the invariable plane of the solar system, but this would introduce conflict with the motion of the nodes of Venus. de Sitter also investigated the motion of the Earth's node. Seeliger's 1906 hypothesis gave $0''.47$ or $1''.13$ without the rotation, or $0''.15$ without the outer ellipsoid. Though Newcomb had not been able to give a value for the discrepancy in this quantity, one is able to determine its effect by looking at the planetary precession which is essentially the motion of the equinox due to the variation in the inclination and nodes of the Earth's orbit. The planetary precession affects the measurements of the right ascensions of stars but not their declinations, so that if one measured the total precession from right ascensions and declinations separately a difference between these two values might indicate a motion in the Earth's orbit that had not been allowed for. This is just what was found (de Sitter 1914, p. 35). Newcomb obtained a difference of $0''.47$, Boss a difference of $0''.85 \pm 0''.22$, and Struve one of $0''.93 \pm 0''.80$. These differences were comparable with the value calculated from Seeliger's hypothesis, i.e. $0''.47$ for the complete version. However, this could not be taken as an unequivocal confirmation of Seeliger's hypothesis, as the discrepancy had already been explained by Hough and Halm (1910) who had shown that this would be produced by a certain unequal distribution of stars in the two Kapteyn star streams.

We can now summarize the empirical consequences of Seeliger's hypothesis as known in 1914. As far as the inner planets were concerned the Newcomb anomalies were all well represented except for the eccentricities which were unimportant. One could not distinguish between Seeliger (1906), de Sitter A or B, or a variation C (de Sitter 1914) which consisted of the inner ellipsoid and the rotation ($6''.85$) only (Table 4.5). The motion of the Moon was not

TABLE 4.5

	Mercury	Venus	Earth	Mars
$e\,d\varpi/dt$	−0.02	−0.012	−0.04	0.00
$\sin i\,d\Omega/dt$	−0.31	+0.05	—	−0.24
di/dt	−0.15	+0.23	−0.17	−0.01
Values added in 1914 for Table 4.4 for earth				
Seeliger	+0.28	A	+1.18	

All values are in units of arcseconds per century.
From de Sitter 1914, p. 35.

sensibly disturbed by the assumed amount of matter. The motion of the Earth became the fine testing ground for the four hypotheses. The motion of the ecliptic was found to be small, but the rotation introduced by Anding (7″.3) and adopted by Seeliger (5″.85) was required in smaller measure (4″.61, the decrease due to the changed constant of precession used by de Sitter) to counter the effect of the outer ellipsoid. This in turn was needed for the advance of the node of Venus. The matter produced an effect on the nodes of the Earth's orbit which appeared to be verified in the discrepancies in precession measurements, but these could equally be explained by unequal star distributions.

There is one area of evidence that has so far been neglected, and that is the evidence of direct observation. Seeliger's hypothesis is often referred to as the 'zodiacal light hypothesis' but it is becoming apparent that this is not the correct description. This view was taken by Newcomb when, in his last paper published posthumously, he withdrew his opposition to the matter hypotheses and accepted Seeliger's hypothesis. He added that 'the matter of Seeliger's hypothesis is not really identical with that of the zodiacal light because it is too near the sun to be visible' (Newcomb 1912). Now Newcomb's concentration on the perihelion of Mercury had led him to neglect the outer ellipsoid which extended to just beyond the Earth's orbit. This was the matter that was meant to be responsible for the zodiacal light, and its existence was thus directly optically confirmed. Its gravitational importance was minor for its main consequence was the advance in the nodes of Venus. The inner ellipsoid was responsible mainly for the highly important advance in the perihelion of Mercury but its semi-major axis was 0.24. Now this was not that close to the Sun, being more than half the distance of Mercury, but de Sitter joined Newcomb in declaring it to be invisible. There was a certain conflict in de Sitter's mind: on the one hand he declared that Seeliger's hypothesis was 'nothing more nor less than a determination of the mass of a material body whose existence was known beforehand' (de Sitter 1913*b*, p. 302), and on the other hand that the greater part of that mass was held to be invisible (de Sitter 1914, p. 33). What was needed, and what would in fact come soon, was a discussion of the visibility of clouds of material particles and whether the mass assigned to Seeliger's ellipsoids was compatible with their visibility.

There was a third component of the original hypothesis of Seeliger. This was the rotation of the fixed star system with respect to the planetary inertial system and was the most improbable part of the 1906 hypothesis. It did appear in 1913 that de Sitter could make this rotation an unnecessary addition, but it had to be reintroduced in 1914. This return was vital, to use de Sitter's adjective, because without it the discrepancy in the secular variation of the inclination of the ecliptic was far too large. This was partly due to the high inclination of the outer ellipsoid, and this high inclination was in turn needed to account for the advance in the nodes of Venus. If one had regarded the introduction of the rotation as decreasing the probability of Seeliger's hypothesis, and this seems to have been generally accepted, then the need to account for the nodes of Venus involved making Seeliger's hypothesis improbable. In Chapter 3 it was argued that the need to account for the same anomaly had led Newcomb's use of Hall's hypothesis to be improbable. The anomaly in the nodes of Venus had now become a sticking point for both approaches to the solution of Mercury's perihelion advance. The astronomers with their faith in the accuracy of their work hoped to solve all Newcomb anomalies in the motions of the inner planets. The physicists had no responsibility for describing the planetary orbits and they could afford to ignore the lesser anomalies, especially the awkward nodes of Venus. In particular Einstein ignored it, as he recalled when he wrote to the daughter of Simon Newcomb in 1929:

> Only in the case of one planet was there found a slight deviation from the calculated orbit, a deviation which exceeded the limits traceable to errors in observation; it was the case of Mercury, the planet nearest the sun. (Quoted by Brasch 1929)

Was Einstein ignorant of the other anomalies, or had he judiciously weighed the evidence? Or was this a case of suppressing falsifying evidence, as the nodes of Venus were still anomalous in 1929? It was of course Einstein's general relativity which gave the eventual explanation of the perihelion advance of Mercury. In order to accept it one had to reject Seeliger's hypothesis, and we will consider three scientists, each important in his own way, and the circumstances in which they chose to adopt the relativistic path.

4.6. Freundlich's attack on the hypothesis

Erwin Freundlich was a friend and colleague of Einstein who supplied him with astronomical advice and information in the crucial years leading up to general relativity. Einstein described him as his *Gesprachspartner* in astronomical problems (in a letter dated 15 July, 1915 (Hermann 1968, p. 32)). Freundlich had been appointed an observer at the Berlin Observatory in 1910 at the age of 25, having just finished his Ph.D. under Felix Klein at Göttingen. He met Einstein in 1911 and in 1918 started work full time for him. In 1914 he had organized an eclipse expedition which went to the Crimea in order to test the Einstein predictions on the deflection of starlight in the Sun's gravitational field. The expedition was stopped by the outbreak of World War I. Freundlich was captured by the Russians but was soon freed in an exchange of prisoners. Frustrated in his attempt to test starlight deflections, Freundlich now turned his attention to perihelion advances. In early 1915 Einstein still had some way to go before completing general relativity which appeared in the November. The field equations of early 1915 predicted a perihelion advance of 18″ per century, as Einstein wrote in a letter to Sommerfeld (Hermann 1968, p. 32). If

the new theory was to be accepted, any rival explanation of the existing perihelion advance of Mercury had to be refuted. Accordingly Freundlich set to work on Seeliger's hypothesis, and on 27 February 1915 he published an attack on it.

The main thrust of the paper concerned the visibility of the zodiacal light matter, but Freundlich also criticized the rotation *r*, the small effect of the third (outer) ellipsoid, and the orbitrariness of the mass distribution. He used a result of Hirn to show that a medium of density 4 g km^{-3}, which Seeliger had assigned to the outer ellipsoid, would shorten the periods of revolution of the inner planets by sensible amounts and that to agree with the Earth's motion the density had to be about a thousandth of this. This had been calculated for a stationary medium and the resistive effect of the small particles would be much less if they were moving round the Sun under gravitational attraction. However, one had to take into account for such small particles the effect of light pressure and it was possible that the zodiacal light particles would then have to be much larger ('like minor planets'). Direct evidence such as shooting stars, comet tails, and the solar corona tended to tell against the existence of large particle sizes.

A second approach was then developed. The volume of the zodiacal light matter, stretching beyond the Earth's orbit, was $2 \times 10^{21} \text{ km}^3$. Freundlich made the assumption that the particles had a cross-sectional area of 1 mm^2, were of similar albedo to Mercury, and of individual density 3 g cm^{-3}. Taking the brightness of the zodiacal light as equal to that of a star of first magnitude, he estimated that there were 9×10^{18} particles† so that the mean density of the matter was $7 \times 10^{-6} \text{ g km}^{-3}$. This was much less than the value required by Seeliger and supported the conclusion that the matter could have only insensible gravitational influence on the planetary motions. In addition this value seemed to be supported in an independent manner, for Arrhenius had estimated the mass of the solar corona to be 12×10^9 kg giving a density of about $1.2 \times 10^{-6} \text{ g km}^{-3}$. This was of the same order as the density of the zodiacal light matter which Freundlich had estimated and he rightly felt that this justified his value. He conceived of the corona as a finely distributed mass whose gravitational attraction towards the Sun was balanced by light pressure. Seeliger's inner ellipsoid was no near the Sun that it must have had some connection with the corona and one would have thought that it was improbable for the overall mass distribution, in general decreasing with distance from the Sun, to increase suddenly by a million-fold.

Freundlich's paper was not considered by Einstein as having been one of his major services. Einstein referred to Freundlich's refutation of Seeliger's theory in a letter dated 2 February, 1916 (Hermann 1968, p. 39), where he described it as 'forcing an open door' (*einrennen einer offene Thur*). But the zodiacal light hypothesis was not 'an open door' for all astronomers.

† This is much the same as the value quoted by Russell *et al.* (1928).

In September of 1915 Seeliger published his own visibility relations (Seeliger 1915). He did not agree that the particles forming the zodiacal light matter should be restricted in size to the order of millimetres and took for his example 1 m (even instancing 50 m). Freundlich's procedure was reversed and a brightness distribution was derived from the 1906 density value for the inner ellipsoid—the light was half as bright near the Sun as the surface brightness of the Moon, decreased inversely with the distance, and disappeared at about 17.5° from the Sun (the maximum displacement of Mercury is 28°). This had little in common with the zodiacal light. However, it also seemed that this had little in common with the light distribution near the Sun. Turner (1900) had measured this during the 1898 eclipse and found that near the Sun the brightness was 2.2 times the luminosity of the full Moon but that it decreased as the inverse sixth power of the distance. This was the coronal light, but the other term was a constant 0.012 times the luminosity of the full Moon and was due to the general luminosity of the sky and the zodiacal light. Now the Sun's angular radius is 16′, so that Seeliger's 17.5° is about 70 Sun radii off centre. At an observed brightness decrease of the inverse sixth power the observed cut-off was going to occur long before Seeliger's distance was reached. Turner's results were reputable—Harold Jeffreys used them in 1919 and they had been published in 1900 so they were freely available. The evidence then did seem to go against Seeliger. In summary, Freundlich had shown that the zodiacal light matter was too sparse to affect the planetary motions and that Arrhenius's measurement of the coronal density was similar to the zodiacal light density and thus also too fine, even more so as Seeliger had demanded a higher density for the inner ellipsoid. Seeliger's own brightness distribution was inconsistent with the observed one.

4.7. De Sitter's abandonment of the hypothesis

The second person to be considered did not abandon Seeliger's hypothesis on visibility grounds. This was de Sitter, whom we have already seen as a proponent of the hypothesis. de Sitter was professor of astronomy at the University of Leiden in Holland and was influential in bringing general relativity to the attention of the English speaking world, a role which Eddington then took up. de Sitter published three long papers in the *Monthly Notices of the Royal Astronomical Society* in 1916 and 1917, entitled *On Einstein's theory of gravitation, and its astronomical consequences* (de Sitter 1916*a*,*b*, 1917). At the end of the first paper de Sitter discussed the consequences for planetary motion. As is well known, the only effect was a perihelion advance but in retaining his new constant of precession de Sitter also gave extra advances to the perihelia and nodes. The differences between theory and observation are shown in Table 4.6. Of these, the perihelia of Mercury and Mars were still discrepant, but only slightly, and the advance in

TABLE 4.6

		Observed	Theory	Difference	Difference as given by Newcomb
Mercury	$e\,d\varpi$	+118.00	+118.58	−0.58±0.43	+8.48
	$\sin i\,d\Omega$	−92.04	−92.50	+0.46 0.52	+0.61
Venus	$e\,d\varpi$	+0.28	+0.39	−0.11 0.25	−0.05
	$\sin i\,d\Omega$	−105.47	−106.00	+0.53 0.17	+0.60
Earth	$e\,d\varpi$	+19.46	+19.46	0.00 0.13	+0.10
Mars	$e\,d\varpi$	+149.44	+148.93	+0.51 0.35	+0.75
	$\sin i\,d\Omega$	−72.64	−72.63	−0.01 0.22	+0.03

All values are in units of arcseconds per century. From de Sitter 1916a, p. 728.

the nodes of Venus was about three times its mean error. This was now an anomaly for general relativity but it was never taken as such in the same way that the perihelion of Mercury was for Newtonian theory. This was not because of the magnitude of the anomaly as much as the impossibility of explaining the anomaly on its own. de Sitter just left it at this stage. He now argued that Seeliger's hypothesis was 'superfluous'. This was at least consistent with his earlier position for in 1913 he had described the action of this hypothesis as simply determining the mass of the zodiacal light matter from the planetary anomalies. If a prior commitment to Einstein's theory had convinced him that there were no significant anomalies remaining then he could say that the perturbing mass was insignificant. Furthermore, his commitment to general relativity could be based on its *a priori* structure rather than on its remaining empirical consequence which were at this time either equivocal (stellar red-shift) or not yet tested (deflection of starlight). He should have given an argument that the observed brightness of the zodiacal light matter was capable of being produced by a gravitationally negligible amount of matter, but of course he had not quantitatively discussed this aspect before. However, I doubt if he would have been troubled by this for in 1914 he had declared the inner ellipsoid to be invisible even with a large mass and had not been particularly keen on retaining the outer ellipsoid. The omission is a sign of the rather *laissez faire* attitude to Seeliger's hypothesis.

4.8. Jeffrey's rejection of the hypothesis

The third person whose attack on Seeliger's hypothesis will be considered is Sir Harold Jeffreys. Jeffreys is interesting in that he was a late proponent of the hypothesis and only abandoned it in 1919. It was the early stages of his career. He had been elected a fellow of St John's College, Cambridge in 1914, was

Isaac Newton student from 1914–1917, and worked at the Meteorological Office in London from 1917–1922. Jeffrey's first paper on the zodiacal light hypothesis was published in 1916, in the same issue of the *Monthly Notices* as de Sitter's second paper on general relativity (Jeffreys 1916). He noted that the motion of the nodes of Venus was not explained by Einstein and on that account reinvestigated Seeliger's hypothesis. He first rejected the rotation as 'too great to be possible' and cited de Sitter's 1913 article in support. However, de Sitter's 1914 article had shown that the rotation could not be abandoned, and it is probable that owing to the war Jeffreys had not been able to see that paper. Fortunately this did not matter, for Jeffreys rejected the outer ellipsoid (which was responsible for the zodiacal light) and it had been the effect of this on the Earth's motion that had to be countered by retaining the rotation. The rejection of the outer ellipsoid as a cause of the planetary anomalies was based on a visibility argument. He showed that the diameters of the reflecting particles had to be of the order of 10^7 cm to fit the observed brightness and the density required by Seeliger; at this size of course they would appear as minor planets. On the other hand if the particles were small compared with the wavelength of light the light would be scattered rather than reflected, and at the observed brightness the density of the zodiacal light matter could not in any way be sufficient to perturb the motions of the inner planets sensibly. This by now is a familiar approach for us though the bringing in of scattering was novel. Since there were no references to Freundlich, the 1915 Seeliger paper, or the Dutch papers, I presume that these were unavailable during the war so that Jeffreys's approach was independently worked out.

The theory was then developed for the perturbative effects of an oblate distribution of matter close to the Sun. The comparison of the theory with Newcomb's anomalies is given in Table 4.7. The only remaining anomalies (i.e. those quantities where the difference between Jeffreys's variation and the corresponding Newcomb anomaly was greater than the mean error) were in the perihelion of Mars and the nodes of Venus. The former was still twice its mean error—a mass distribution well inside the orbit of Mercury was going to

TABLE 4.7

Planet	$\sin i\, \delta\Omega$	δi	$e\, \delta\varpi$
Mercury	0.94	0.00	8.53
Venus	0.41	0.42	0.04
Earth	—	—	0.03
Mars	0.04	0.00	0.03

All values are in units of arc seconds per century. From Jeffreys 1916, p. 118.

have little effect on Mars. Jeffreys found no trouble in the invisibility of this matter:

> Within the orbit of Mercury the observations are not so decisive, as they must be made so soon after sunset or before sunrise that matter there is practically invisible. (Jeffreys 1916 p. 113)

Now this was being lazy, for there was a well-known distribution of matter near the Sun, i.e. the solar corona. Estimates were available for its brightness, and even if it was not taken to be the perturbing matter some sort of reason should have been given why the more massive perturbing matter was not visible. Without it the hypothesis was incomplete. Furthermore, since the ellipsoid had an inclination of 8° 19′ it was subject to Newcomb's objection to high inclinations. Jeffreys did not mention this—there could possibly have been a connection with the Sun's equator which has an inclination of almost 8°.

As far as the planetary anomalies were concerned Jeffreys's discrepancy for the perihelion of Mars was pitted against Einstein's discrepancy for the nodes of Venus. The former was twice its mean error; the latter three times its respective mean error. The perihelion value was not only less discrepant, it would be easier to account for in other ways. In these papers Jeffreys paid no attention to *a priori* aspects of gravitational theories. He concluded the 1916 paper:

> It appears therefore that on the hypothesis of Einstein it will be found very difficult to find a plausible explanation of the motions of the planes of the orbits, and in particular of the node of Venus. (Jeffreys 1916, p. 118)

In 1918 Jeffreys applied Seeliger's hypothesis to the case of Silberstein's relativistic theory of gravitation which had been put forward earlier in 1918 (Silberstein 1918). Silberstein supported Einstein's requirement of general covariance but attacked the equivalence hypothesis. This was partly on methodological grounds—for instance it placed gravitation 'on an entirely exceptional and privileged footing'—and partly on empirical grounds. Einstein's theory predicted a red-shift in stellar spectral lines but the verification of this effect was proving difficult. In 1914 Freundlich had found the required positive effect in stellar lines, but in 1917 St John had found a negative result in the solar spectrum. Silberstein preferred St John's data—they were more numerous, less dubious, and the solar mass was better known than Freundlich's stellar masses. Furthermore, an additional effect of the deflection of starlight in the Sun's gravitational field had yet to be tested. Now if there was no Einstein red-shift there could be no Einstein perihelion advance, and this apparent success had to be discounted. Silberstein effectively used St John's result on the red-shift to discredit the equivalence hypothesis and his own theory used just the general covariance requirement (though here is not the place to discuss the theory (see Chapter 7) one might use Kretschmann's result† to ask whether this was so significant). Instead of an advance in the perihelia, Silberstein's theory predicted a regression. These were quite small, the values of $ed\varpi$ for Mercury, Venus, Earth, and Mars being $-1''.48$, $-0''.010$, $-0''.011$, and $-0''.21$ respectively. This increased the anomaly for Mercury from $8''.48$ to $9''.96$ but he was not perturbed:

I understand from a conversation with Mr. Harold Jeffreys, who has already found a satisfactory representation of Mercury's $8''.48$ and of the motion of the node of Venus by means of a modification of Seeliger's zodiacal-light matter that the above, increased, excess of about 10% could equally well, and possibly 'more easily', be accounted for by an appropriate distribution of the said disturbing matter. (Silberstein 1918, p. 128)

Jeffreys's paper was published a little later in 1918 (Jeffreys 1918). The disturbing matter was similar in distribution to the 1916 matter but of greater density. The main discrepancy remaining was, as before, the perihelion of Mars. Jeffreys suggested that this could possibly be referred to an increase in the mass of the Earth or to the asteroids. The effect on the Earth's orbit would be within the probable errors. This was not inconsistent with de Sitter's 1914 result because it had been the outer ellipsoid that had caused trouble for the Earth. Jeffreys finished:

We can conclude herefore that Dr. Silberstein's theory, combined with gravitating matter near the Sun, can give an excellent representation of the secular inequalities of the inner planets. (Jeffreys 1918, p. 205)

Jeffreys's retraction of Seeliger's hypothesis came in 1919 (Jeffreys 1919). In this year the Eddington eclipse expedition had confirmed Einstein's prediction of the deflection of starlight in the Sun's gravitational field. Since the red-shift results were equivocal and the perihelion advances could be explained in other ways (at least apparently), Eddington's positive results were powerful evidence for Einstein. There was the possibility that the starlight deflections could have been caused by refraction in the solar corona, and this forced attention to be turned to the constitution of the corona. In particular the density had to be found. The move that Jeffreys made in 1919 was to identify his perturbing matter—previously 'invisible'—with the coronal matter. Then declaring that 'you can determine the density perfectly from the luminosity'‡ he proceeded to do so. This was used to show that any refraction of star images in the corona was practically insensible. Einstein's theory had now a definite advantage over the classical theory on this point. On empirical grounds it had to be taken seriously, and, since it predicted a perihelion advance for Mercury that was almost exactly that observed, this advance could no longer be taken

† All physical laws could be written in a generally convariant form (Kretschmann 1917).

‡ This was stated by Jeffreys during the meeting of the Royal Astronomical Society of 12 December 1919, which was devoted to a discussion of the theory of relativity. Those speaking included Eddington, Jeans, Sir Oliver Lodge, Jeffreys, Lindemann, and Silberstein. The discussion is to be found in (Fowler 1919). For Jeffreys's remark see p. 116.

as an anomaly with complete certainty. In this case one could not use the perihelion advance to support the existence of invisible matter, and if its existence was to remain acceptable independent evidence had to be provided. This could only be optical evidence, which ruled out invisible matter and in this case allowed just the zodiacal light matter and the coronal matter. In 1919 he re-investigated both. It was shown that the zodiacal light matter, if in the form of gas molecules, would be gravitationally insensible, and, if in the form of small solid particles, could produce at most 2″ in the centennial advance of Mercury's perihelion. The corona was assumed to be due to sunlight scattered by diffuse matter, and using Turner's 1898 observations Jeffreys showed that this matter could have only an insignificant disturbing effect on the motions of the planets. Seeliger's hypothesis had been laid to rest.

4.9. The advance in the nodes of Venus

The critical anomaly which the new general relativity failed to explain was the motion of the nodes of Venus. The wish to explain it had led Newcomb into trouble with solar parallax determinations. It had led Seeliger to reject Hall's hypothesis and Jeffreys to adopt Seeliger's hypothesis. In 1919 Jeffreys found that this zodiacal light hypothesis could not be used to explain the planetary anomalies that Newcomb had discovered in 1895. Faced with this he concluded that the advance in the nodes of Venus was not a real effect, but somehow had been obtained in error. Newcomb had tentatively suggested this at times, but Jeffreys was much more positive in tone (Jeffreys 1919, p. 141–2):

> Thus neither the zodiacal matter nor the corona is quantitatively competent to affect Mercury and Venus sufficiently to furnish any argument against Einstein's theory. The excess motion of the node of Venus, if confirmed by future observations, cannot then be accounted for by the attraction of intra-Mercurial matter; at the same time, now that such matter is known not to exist in sufficient quantities to affect these planets, the fact that the excess motion of the node of Venus requires a considerable quantity to account for it affords an argument against the reality of this excess motion, but none against the theory of Einstein; for this and the Newtonian theory alike are now shown to require such a motion should not exist.

Jeffrey's conclusion was indicated in 1958 when Duncombe published a theory of Venus that was to become the successor to that of Newcomb. He found no anomaly in the node:

> The large discrepancy between the observed and theoretical motion of the nodes of Venus found by Newcomb is not confirmed here. The reason for Newcomb's discrepancy has eluded explanation but is probably attributable to large systematic errors in the older observations. (Duncombe 1958, p. 44)

It is somewhat of an anticlimax to find that this anomaly, so important as a companion to that of the perihelion of Mercury, came to such an unimposing end.

4.10 Charles Lane Poor

A discussion of Seeliger's hypothesis should not end in 1919 for opposition to Einstein continued for some years. Much of this had personal and racial undertones, but some of it was due to genuine disquiet about general relativity. Ernst Grossmann questioned the rigour of Newcomb's work (Grossmann 1921). He thought that the perihelion advance for Mercury should be between 29″ and 38″ and that this was too small to confirm Einstein's theory. Other astronomers accepted all of Newcomb's anomalies and wished to explain them in full. Charles Lane Poor, Professor of Astronomy at Columbia University, believed in the efficacy of Seeliger's hypothesis:

> Einstein and his followers have cited the motions of the planets as proof of the truth of his hypotheses. The evidence does not sustain this—his hypotheses and formulas are neither *sufficient* nor *necessary* to explain the discordances in these motions. They are not *sufficient*, for they account for only one among the numerous discordances—that of the perihelion of Mercury; they are not *necessary* for all the discordances, including that of Mercury, can readily be accounted for by the action, under the Newtonian law, of matter known to be in the immediate vicinity of the sun and the planets. (Poor 1921, p. 33)

In earlier writings (Poor 1905 and 1908, p. 171) Poor had shown favour to material explanations of the Newcomb anomalies. Now he thought that he was able to account for each of them with suitable matter distributions. In his book *Relativity versus gravitation* (1922) Poor extended his scheme to include the deflection of starlight by matter surrounding the Sun. Elsewhere (Poor 1925, 1930) he gave certain worthless theoretical criticisms of general relativity. Nowhere did Poor discuss the quantitative aspects of the visibility of his disturbing matter. Harold Jeffreys (1921) pointed this out to him in a reply to the 1921 paper quoted above. Poor did not take in these limitations, and by not doing so, forfeits our attention.

5

SOME THEORETICAL ARGUMENTS FOR THE INVERSE SQUARE LAW

5.1. Gravitational laws

So far in this book we have been concerned with the linear development of the history of the anomalous advance in the perihelion of Mercury. An account has been given of the discovery of the anomaly, how astronomers tried to explain it, and how the zodiacal light hypothesis was taken as a satisfactory explanation until general relativity superseded Newtonian gravitational theory. It is now necessary to consider the background to the arrival of Einstein's theory.

So far astronomers have played the leading roles. Now physicists predominate, particularly those who were interested in gravitation. In Chapter 1 it was shown that a perihelion advance of a planet orbiting a central body could be produced either by disturbing matter obeying the Newtonian gravitational law or by the gravitational interaction between the planet and the central body being governed by a law different from the Newtonian inverse square law. Apart from Hall's hypothesis, described in Chapter 3, the search for non-Newtonian laws involved the search for a deeper meaning to gravitation including explanatory mechanisms for it. The quest for such mechanisms had gone on for centuries and Newton, at least in public, had emphatically dissociated himself from that task in the *Principia* when he wrote the celebrated phrase 'hypotheses non fingo' ('I frame no hypotheses'—Newton 1934, p. 547) about the cause of gravity. But scientists attempted with renewed fervour to explain the inverse square law which Newton had successfully introduced. In 1872 Mach reflected on this:

> The Newtonian theory of gravitation, on its appearance, disturbed almost all investigators of nature because it was founded on an uncommon unintelligibility. People tried to reduce gravitation to pressure and impact. At the present day gravitation no longer disturbs anybody: it has become *common* unintelligibility. (Mach 1911, p. 56)

If scientists were no longer disturbed, there were at least some who still wished to provide the inverse square law with a theoretical underpinning. In a review article Taylor (1876) described twenty-one theories purporting to do this, a list that was extended in 1881 by Stallo (1960) and then later by Zenneck (1903). Three nineteenth century theories will be considered in detail below; they involve atomic pulsations transmitted through the ether, forces between ether 'sources' and 'sinks', and innumerable tiny particles flying through space bombarding bodies. At the same time as these were appearing the new branch of physics of electrodynamics was being developed. It had an immediate effect on gravitational theories as a number of physicists looked at the possibility of extending the inverse square law in the same way as Coulomb's law had been extended in electrodynamics. The rise of electrodynamics had more long term effects. Some ambitious scientists attempted to produce a single grand scheme which would give both electrodynamic and gravitational forces at once and possibly everything else as well. More importantly, special relativity was introduced as a contribution to electrodynamics in 1905, and quickly set a standard for other theories including gravitation. Could the Newtonian law be reconciled with the principle of relativity? Could special relativity be generalized to produce a gravitational theory in its own right? Such questions produced a number of alternative theories in the years between 1905 and 1915 including, of course, general relativity.

The new gravitational theories can be classified to a certain extent by the way in which they alter the original Newtonian law, in which the gravitational force F between two bodies of mass M and m a distance r apart is, including a gravitational constant G:

$$F = GMm/r^2 \tag{5.1}$$

A slight alteration to the distance dependence gives:

$$F = GMm/r^{2+\delta} \tag{5.2}$$

where δ is a number much less than unity. This is known as Hall's hypothesis, and Simon Newcomb used it in 1895 with $\delta = 0.000\,000\,16120$. This law was discussed in Chapter 2. The inverse square law could also be altered to give:

$$F = GMm\,(1/r^2 + f(r)) \tag{5.3}$$

where $f(r)$ could be typically a/r^4 for some small constant a. This is called Clairaut's law and is discussed in this chapter. A more radical alternative to the inverse square law would be:

$$F = GMm\,(1/r^2 + f(r, \mathrm{r})). \tag{5.4}$$

In this case the force would depend on the distance between bodies and on their velocities; laws of this type arose as analogies to similar laws in electrodynamics. They will be considered in detail in Chapter 6.

Theories which were direct rivals to general relativity are considered in Chapter 7, where their empirical results for planetary perihelia are highlighted. General relativity itself does not retain 'force' as a fundamental concept. It considers the motions of particles in space in terms of how space (or space-time) is distorted in the vicinity of massive bodies. The development of general relativity is discussed by relating each stage to empirical results, in particular perihelion motions, in terms of the space-time metric given by the evolving theory. But before that stage is reached we have to look at the earlier scientists who tried to change the inverse square law, at those who tried to support it, and at first, at the instigator of an early crisis in the history of Newtonian gravitational theory—Clairaut himself.

5.2. Clairaut and the lunar apogee: 1747†

It is worthwhile to look briefly at the Clairaut episode, not only because of his hypothesis but also because he found himself in a similar position to Simon Newcomb. Newcomb used a new law to explain the motion of Mercury's perihelion: in 1747 Clairaut had used a new law to explain the motion of the lunar apogee. The motion of the apogee had been treated by Newton in the *Principia* when he wrote that its motion of 3° per revolution could be covered by an $r^{-2\frac{4}{243}}$ power law (Newton 1934, p. 146). He was able to retain the inverse square law by making the Sun's disturbing force wholly responsible for this motion, but Whiteside (1970, p. 130, 1976) has shown that in the then unpublished papers on this result Newton had cooked the proof. That the Newtonian theory apparently gave only half the observed motion of 3° was discovered independently in 1747 by Clairaut, Euler, and d'Alembert.

Clairaut obtained this result having given to the calculation 'all the accuracy that it demanded' (Clairaut 1745, p. 336). Rather than abandon attraction, with its explanations of Kepler's laws, the nodes of the Moon, tidal theory etc., but despite the difficulties of Bouguer and la Condamine in the theory of the figure of the Earth (pendulum experiments failed to agree with determinations of the meridian arc), Clairaut put forward a slight modification of the inverse square law. This hypothesis was of the form

$$F = GMm\left(\frac{1}{r^2} + \frac{a}{r^4}\right) \tag{5.5}$$

where a is a small constant. Newton's formula (see p. 13), giving the effect on the line of apsides of general force laws, can be used to show that this particular law gives for the angle between successive apsides the quantity $180°\{1+a)/(1-a)\}^{\frac{1}{2}}$. Units are chosen so that $r = 1$ for the maximum distance between the body and the centre of force. As a consequence, despite appearances, the motion of the apsides is dependent on the distance.

Clairaut reasoned that since the Moon–Earth distance was about a hundredth of the Mercury–Sun distance, the additional inverse fourth-power term could be sensible as far as the Moon was concerned but insensible as far as Mercury was concerned (it would be a factor of 1/10 000 more feeble).

† Perhaps the best account of the Clairaut episode is to be found in Gautier (1817); see also Waff (1975) and Chandler (1975).

For those worried by any departure from an inverse square law, Clairaut cited other laws differing from that—those governing the roundness of drops of fluid, the rise and fall of liquids in capillaries, cohesion of marbles put in a vacuum, and the bending and refraction of rays of light. He could have added an inverse 5/2 power law for magnetism or the simple inverse power law for caloric molecules discussed by Newton in the *Principia* (cited by Dorling 1974). However, Clairaut realized the need to investigate his law further, and in the following passage anticipated the argument that Simon Newcomb used in 1882 to dismiss this type of hypothesis:

I have only cited the law made up of the square and fourth power of the distance in order to give more easily an idea of my feelings on universal gravitation: it is necessary to enter into analytical details in order to see the inconveniences that this law would have (these inconveniences would be to give far too great an attractive force to contiguous bodies or ones very near to each other, and to make the total gravity on the surface of the Earth too great relatively to that which it is at the distance of the Moon), and the means of forming another which would have been exempt from them. (Clairaut 1745, p. 336)

Later, Clairaut cited the law that Newton had also instanced, i.e. $r^{-2\frac{4}{243}}$, but for those who wished to represent 'the necessary force by a single term' rather than for those who wanted a law that did not suffer from the above difficulties (Clairaut 1745, p. 540).

After his memoir was published, Clairaut was attacked by Buffon and there followed a short controversy.† Buffon's position was generally weak or fallacious. He started by not accepting Clairaut's result and by reaffirming his faith in Newtonian theory, and in this he was at least soon to be justified. Given Clairaut's hypothesis, Buffon preferred to interpret the second term as expressing a magnetic effect and cited a passage of the *Principia* in which Newton had written

In these computations I do not consider the magnetic attraction of the earth, whose quantity is very small and unknown: if this quantity should ever be found out . . . we should then be enabled to bring this calculation to a greater accuracy. (Newton 1934, p. 484)

Clairaut disliked the interpretation of the inverse fourth power term as magnetic in origin, thinking that it would be too small.‡ He suggested a

† G.L.L. de Buffon, Réflexions sur la loi de l'attraction, pp. 493–500; A. Clairaut, Réponse aux reflections de M.de Buffon, sur la loi de l'attraction et sur la mouvement des apsides, pp. 529–48; Buffon, Addition au Mémoire qui a pour titre: Réflexions sur la loi de l'attraction, pp. 551–2; Clairaut, Avertissement de M. Clairaut, au sujet des Mémoires qu'il a donnez en 1747 et 1748, pp. 578–80; Buffon, Second addition au Mémoire . . ., pp. 580–3; Clairaut, 'Résponse au nouveau Mémoire de M.de Buffon', pp. 583–6., *Hist. Mém. Acad.r. Sci.*, **58,** 1745.

‡ Later Poincaré (1953) cited magnetic effects as a possible explanation of the secular acceleration in the Moon's mean motion. The question was investigated by Brown (1950) who concluded they were insufficient.

possible electric term,† but did not disapprove of Bougeur's suggestion that different parts of the Earth attract with different power laws (with a resulting complex overall law of attraction for the Earth–Moon system). However, in 1749 Clairaut discovered that his law of attraction was superfluous. Proceeding with his intention of going into further analytical detail, he took his approximation one step further and found that the theory gave almost the whole motion of the lunar apogee. In announcing his discovery Clairaut kept the reason behind it secret in the hope of winning a prize from the Academy of St Petersbourg (which he did in 1752), but Euler was able to repeat his own calculation and also found that the anomaly was removed. Finally d'Alembert, who was not able to solve the problem until he knew Clairaut's method, was then able to show that the new approximation still left a 30′ discrepancy but that if carried even further full agreement could be obtained (see Gautièr 1817).

5.3. Two arguments for the inverse square law

Whereas Clairaut had found that the apparent lunar anomaly was not in fact anomalous, Newcomb had showed that the Mercury perihelion advance was indeed anomalous and greater than Le Verrier had thought. We have seen in Chapter 3 how Newcomb used Hall's hypothesis to explain this motion.

Discussion of Hall's hypothesis was limited to empirical arguments by a small number of astronomers. Since, as we have also seen, it was refuted by Brown's lunar theory in 1903, its active life lasted only about seven years (though its influence as a theoretical part of Newcomb's tables of the inner planets lasts until today). Newcomb himself seemed wary of taking Hall's hypothesis as expressing the actual law of attraction—his use of 'as if' has been noted when he wrote

> The excess of motion shown by observations in the case of Mercury and Mars, and computed for all four planets as if they gravitated toward the Sun with a force proportional to r^{-n} where $n = 2.000\,000\,161\,20$. (Newcomb 1895, p. 184)

This lack of enthusiasm for Hall's hypothesis can be explained by the fact that no contemporary theory gave a law of attraction of this type. The theories to be discussed below were designed to yield the inverse square law, and though some allowed the possibility of explaining the perihelion advance of Mercury this was not in terms of a Hall-type law. The theories based on electrodynamical analogies which will be discussed in the next chapter also gave perihelion advances, but again not in terms of an inverse $2+\delta$ law. Since these were the only theoretical approaches available at the end of the nineteenth century, these results would have implied a low *a priori* probability

† Electric terms were later suggested to account for the perihelion advance of Mercury. See Gravé and Sokoloff (1926).

for Hall's hypothesis. Furthermore, two additional arguments could have been adduced against this hypothesis.

The first is from the method of dimensions, a method available at the time (see for example Mach 1960). Equality of dimensions in the ordinary inverse square law of gravitation

$$F = \frac{Gm_1m_2}{r^2} \tag{5.6}$$

is ensured by giving to the gravitational constant G the dimensions $M^{-1}L^3T^{-2}$. However, in Hall's hypothesis we appear to introduce a non-integral number of dimensions. Now this is not unusual in science, for instance the velocity of a transverse wave along a stretched string of tension T and of mass m per unit length is given by

$$v = \left(\frac{T}{m}\right)^{\frac{1}{2}}. \tag{5.7}$$

This of course can be written as

$$v^2 = T/m \tag{5.7$'$}$$

so that the formula does not really contain fractional dimensions. A similar transformation of Hall's hypothesis requires us to write

$$F^{100000000} = \frac{\{G'm_1m_2\}^{100000000}}{r^{200000016}} \tag{5.8}$$

where the constant G' does not have the same dimensions as G but the dimensions of which are still integral. However, whereas the formula expressed in (5.7′) is quite acceptable, the 'real' version of Hall's hypothesis expressed in (5.8) can only be described as most unlikely. Escaping this by giving G non-integral dimensions also seems unsatisfactory.

A second argument against Hall's hypothesis is based on one which connects the inverse square law with the three-dimensional nature of space. Consider an attractive body. The intensity of the attraction at any point on the surface of a sphere centred on the body and of radius R has the same value. The surface area of this sphere is $4\pi R^2$. If the total intensity is assumed to be spread equally over the surface of the sphere, the attractive effect will decrease inversely with the square of the distance since a unit area on this surface receives an intensity proportional to $1/4\pi R^2$. In a two-dimensional space the intensity decreases as $1/2\pi R$ and in an n-dimensional space the intensity decreases as R^{1-n}.

This argument was used by Kant to show, on the grounds that the inverse square law of nature was true *a priori*, that space had three dimensions.† Later

† In his *Gedanken von der Wahren Schätzung der lebendigen Kraft*. This and other references are to be found in Jammer (1954).

Ueberweg proved in this way that the noumenal world, the space of things-in-themselves, was also three dimensional.† Now on this argument Hall's hypothesis leads to a space of $3+\delta$ dimensions. The problem with this is that δ is fractional rather than that extra dimensions are being suggested. In the nineteenth century support for extra dimensions came from spiritualists anxious to accommodate manifestations of spiritual life, and the spiritualists numbered a few prominent scientists among them. Some of the ether theories of gravitation had a natural place for an extra dimension in space. However, even these weak indications of the insufficiency of three dimensions could not cope with a fractional dimension and one must conclude that on this argument Hall's hypothesis is unlikely.

Hall's hypothesis was also unlikely on the basis of the gravitational theories under discussion in the nineteenth century. The kinetic theories listed by Taylor (1876) tried to explain gravitation as the result of underlying motions—the motions of innumerable tiny particles or the motions of streams of ether. We will discuss three type of kinetic gravitational theories, the first being the pulsation theories.

5.4. Pulsation theories of gravitation

The central figure in the pulsation 'school' was C.A. Bjerknes, a Norwegian whose hydrodynamical theories tried to embrace electric, magnetic, and gravitational interactions. He began this work in 1856 and published his main papers in the 1870s. Bjerknes's idea was that two spherical bodies immersed in an incompressible fluid and pulsating in phase attracted each other with a force depending on the inverse square of their distance apart. The force was repulsive if the bodies pulsated out of phase by π. Bjerknes found fame with the experiments demonstrating his ideas which he started in 1875 and showed at the Paris Electrical Exhibition in 1881 where he was awarded a Diplôme d'Honneure by the International Jury. The experiments, as described by Cooke (1882), used hollow rubber balls forced to pulsate by an air pump and immersed in water. Cooke carried out experiments of his own using small drums instead of balls and demonstrated analogies to many electric, magnetic, and diamagnetic phenomena. In April 1881 Stroh gave a lecture with demonstrations on Bjerknes's experiments to the Society of Telegraph Engineers and of Electricians in London, a lecture that was so popular that it was repeated in May (Stroh 1882). Stroh referred to another early proponent of pulsation theories, Guthrie (1870), who carried out experiments on the attractive and repulsive forces between vibrating tuning forks. Guthrie in turn referred back to a common source of inspiration for pulsation theories, the French scientist Guyot (1832). Following Stroh's lecture, Forbes spoke of the reaction Guthrie's experiments had caused on their appearance:

† For a discussion see Lange (1881).

When Dr. Guthrie published his experiments on attraction with a tuning fork, some 13 or 14 years ago, many felt that a new world was opened before them, and a sort of expectation existed that the action of gravity, magnetism, or electricity might thereby be explained. (Stroh 1882, p. 224)

In Cambridge a similar expectation spurred on the astronomer Challis to produce a large amount of work on hydrodynamical theories. He achieved little success despite Taylor's description of it as 'the most carefully-studied and the most diligently-elaborated exposition of the wave theory of attraction which has yet been proffered to the scientific world' (Taylor 1876, p. 253). Challis's 'Hydrodynamical Theory of attractive and repulsive forces' of 1872 was left incomplete because he found that the solution of the second approximation contained divergent terms. In 1876 he discussed gravity molecules and gravity waves of very large wavelength. He is of immediate interest in that his 1859 theory gave an additional $1/r^4$ term:

Again, the mathematical theory seems to indicate that, if the approximation were carried further, the law of gravity would be found to be expressed by such a function as $m/r^2 + \mu/r^4$, μ being excessively small. This second term, if at all sensible, would be most likely to be detected in the action of the sun on Mercury. In fact, M. Le Verrier has recently shown that the motion of the perihelion of this planet is not wholly accounted for by the attraction of known bodies. The proper course in this case is no doubt that which M. Le Verrier has recommended, *viz.* to endeavour to ascertain whether there are small bodies circulating between the sun and Mercury, to which the motion of Mercury's perihelion may be ascribed. But should this explanation fail, it might, I think, be reasonably questioned whether the law of gravity is so absolutely that of the inverse square as has been generally assumed. (Challis 1859, p. 450)

Challis added that Bode's law gave no indication of an intra-Mercurial planet; when Lescarbault 'discovered' such a body Challis was sceptical about its planetary nature (Challis 1861, p. 221). The Clairaut law arose because the force law was obtained as a series in which the dominant term was that of the inverse square with extra small (usually very small) terms—this will be seen to be a common feature of these theories. Challis was, I think, the first to suggest that the inverse square law was not exact in this connection, and I do not think it insignificant that he had theoretical backing for this view. This theory used his proof that small spherical bodies were attracted by waves of large breadth and repelled by waves of short breadth.

The successor to Challis was W.M. Hicks who in 1880 published a paper with two main purposes: first to put forward a new method of mass images which 'enables us to determine rigorously the action between two spheres' and second 'to show how this action may be applied to explain gravitation, and especially the gravitation of the vortex atoms of Sir William Thomson' (Hicks 1880, 1880–3). Hicks used Maxwell's argument (see below) to reject Le Sage's theory, which was perhaps rather disingenuous in that it was Thomson who in 1873 had shown how to circumvent that argument. Hicks assumed the vortex atoms to be pulsating in the ether and obtained the following force law:

$$F = \frac{8\pi^3\bar{a}^2\bar{b}\alpha\beta}{\tau^2\, r^2}\cos\left(\frac{2\pi\varepsilon}{T}\right) + \frac{4\pi}{Tr^5}\int_0^T a^3b^4\dot{b}^2\,\mathrm{d}t - \frac{a^3\dot{b}^6}{r^5}. \tag{5.9}$$

Hicks' force is thus a Clairaut hypothesis though he neglected the r^{-5} terms on the grounds that they were extremely small, a and b being the radii of the pulsating particles (i.e. the vortex atoms). The phase difference between the pulsations is represented by ε, T being the period of the pulsations which is the same for all particles. When $\varepsilon = 0$, $\cos(2\pi\varepsilon/T) = 1$ and the force is attractive, but when $\varepsilon = T/2$, $\cos(2\pi\varepsilon/T) = -1$ and the force is repulsive. The force is zero when $\varepsilon = T/4$ or $\varepsilon = 3T/4$. The theory thus allowed 'negative' matter which repelled ordinary matter ('positive' matter) but attracted other negative matter. This was not implausible for any such material would have been expelled from our solar system and taken up a position in a negative solar system. The negative system would look the same to us. The weightless matter could exist in either system, but since the systems would probably be in translatory motion this weightless matter would be left behind. There is of course a lot of matter scattered throughout space, though it is to be remembered that the two types of weightless matter attract or repel each other.

This negative matter is not the same as is possible in Newtonian theory. This was mentioned in Chapter 4 in regard to the gravitational equivalent of Olbers' paradox and the suggestions of Föppl and Schuster. One piece of evidence in need of this type of support was the high velocity of the star Groombridge 1830 noticed by Simon Newcomb. On Newtonian theory positive matter attracts all matter because the mass of the attracted body cancels out owing to the quality of inertial and gravitational mass. Similarly negative matter repels all matter. Now consider a negative body following a moving positive body. The latter attracts the former but is itself repelled by it. The positive body therefore accelerates in the direction in which it is already moving, as does the negative mass. Eventually speeds will be attained that are higher than one would expect if there were just positive masses moving under mutual attractions, and taking one of the bodies as the star Groombridge 1830 with a dark companion of the opposite type one has the possibility of explaining that star's excessive speed. However, the type of negative mass involved here will not give this sort of effect, and Schuster was wrong in thinking that they would (he cited Pearson, for which see below).

In 1889 the problem was taken up by Leahy who differed from Hicks in that he treated the pulsations as occurring in an elastic medium rather than one which was incompressible (Leahy 1889). He obtained for the force between two pulsating spheres a distance c apart the expression:

$$F = -\frac{24\pi^2\mu(\lambda+2\mu)}{\rho(2\lambda+5\mu)}\rho_1\rho_2\frac{b^3a}{c^2} - \frac{72\pi^2\mu(\lambda+2\mu)^2}{\rho(2\lambda+5\mu)^2}\rho_1^2\frac{ba^4}{c^3} \tag{5.10}$$

where a and b are the radii of the two spheres, λ and μ are coefficients of elasticity, ρ frequency of pulsations, ρ_1 and ρ_2 are constants for the two spheres representing the amplitudes of pulsations, i.e.

$$A^{\frac{1}{2}} = a(1+\rho_1 \sin \rho t).$$

The inverse cube term is always repulsive, whilst the inverse square term is repulsive when ρ_1 and ρ_2 have the same sign and attractive when they have different signs. Leahy's force is thus a Clairaut hypothesis. However, to be attractive the force requires the two bodies to be pulsating out of phase and this is a distinct objection. Whereas any number of bodies can pulsate in phase, only two bodies can pulsate mutually out of phase. If we have two pulsating spheres with an attractive force between them, then on Leahy's theory a third pulsating sphere will attract one of the pair and repel the other. Since these spheres are meant to be the ultimate particles of matter, the building up of a large body will provide insuperable difficulties unless strong cohesive forces are imported.

Since Leahy could not conceive of an absolutely incompressible ether he would not have abandoned its elasticity. Moreover, the ether as used in electromagnetism was elastic, and to have to coextensive ethers of opposite properties seemed to him to be unlikely—for this reason he criticized the theory of Hicks. He also objected that at distances greater than a quarter-wavelength attraction would become repulsion, but since the law of attraction had not been verified for distances greater than the Neptune–Sun distance one just required that the quarter-wavelength was greater than this. Leahy did not give any values for this, but did give values for the electrostatic case in which the wavelength was 200 miles. It would be difficult to test Coulomb's law at a distance of 50 miles.

In Leahy's force law the ratio of the r^{-3} term to the r^{-5} term is equal to

$$\begin{aligned} &\frac{3(\lambda+2\mu)a^3}{(2\lambda+5\mu)b^2}, \text{ assuming } \rho_1 = \rho_2 \\ &= \{3(\lambda+2\mu)/(2\lambda+5\mu)\}a, \text{ assuming } a = b \\ &\approx a. \end{aligned}$$

Thus Leahy's force is

$$F = -\text{const.}\left(\frac{1}{r^2}-\frac{a}{r^3}\right). \qquad (5.10')$$

Since a is the radius of an atomic particle it is going to be small, and the apsides will regress a very small amount each revolution.

Leahy's theory is obviously of little practical use. In 1895 Hicks again supported pulsation theories of gravitation in a Presidential Address to the Mathematics and Physical Science Section of the British Association, though his confidence was reflected in his calling it the 'least unsatisfactory' of such theories (Hicks 1895). In 1906 and 1907 Poincaré devoted a section of his lectures on *Les limites de la loi de Newton* to pulsation theories (Poincaré 1953, Chap. 15, pp. 257–65). Not surprisingly he was not impressed and he gave two difficulties of the theories: firstly the requirement that all the atoms had to pulsate in the same phase seemed to need an explanation of why they did this; and second external work had continually to be supplied to keep the amplitude of pulsation the same. Poincaré noted that the 1898 theory of Korn met the first requirement by supposing that the fundamental frequency of a system of vibrating particles was predominant and by attributing gravitation to that fundamental frequency. However, Poincaré found various difficulties in Korn's theory.

The English school of pulsation theories reached its peak with C.V. Burton. In 1909 he put forward a theory which also answered Poincaré's first difficulty by suggesting that compressional–rarefactional waves were being continually propagated through the ether (1909). As opposed to the theory of Challis, to which Burton referred and in which these waves themselves caused bodies to be attracted, Burton's waves caused the fundamental particles (electrons) to pulsate thereby causing attraction. The ether was taken as nearly incompressible, the waves being propagated at very high velocity, but the 'greater universe' acting as the source of the waves was beyond our knowledge. Burton obtained a force of the form

$$F = r^{-2}\cos(2\pi r/\lambda) \tag{5.11}$$

in which λ is the wavelength of the ether waves. Actually this was for two bodies whose line of centres was perpendicular to the direction of propagation of the waves, and one really needs waves travelling equally in all directions which presumably would involve interference. Expanding as the first few terms of a power series gives

$$F = \frac{1}{r^2} - \frac{\pi^2}{2\lambda^2} + \frac{3\pi^4}{4\lambda^4}r^2 - \ldots. \tag{5.11$'$}$$

The law is thus a Clairaut hypothesis. Using the Newton formula for the effect on the line of apsides of such a force gives for the angle between successive perihelia

$$2\pi\left(\frac{1+3\pi^4/4\lambda^{4'}}{1+5\times 3\pi^4/4\lambda^4}\right)^{\frac{1}{2}} \approx 2\pi\left(1+\frac{3\pi^4}{8\lambda^4}\right)\left(1-\frac{15}{8}\frac{\pi^4}{\lambda^4}\right)$$
$$\approx 2\pi\left(1-\frac{3\pi^4}{2\lambda^4}\right)$$

which is a very slight regress. Attraction changes to repulsion at a distance $\lambda/4$, so observation requires $\lambda/4 > 30$ AU. Burton suggested $\lambda = 10^4$ AU, which means that departures from the inverse square law are negligible. Furthermore the corresponding velocity of the ether waves is 2.19×10^{27} cm s^{-1}, which is high enough to avoid aberration effects.

In his *Treatise on electrical theory* de Tunzelmann (1910, p. 439) suggested that the anomalous perihelion advances might be explained by the existence of a non-electromagnetic term in the inertial mass of an electron. This gravitational term was a function of the mass of ether extruded by an electron during pulsation, and on the deformable (non-constant volume) model of the electron one would expect this to vary with velocity. Nevertheless the term is extremely small, being given by $5/3 \times c \times$ mass of ether extruded by an electron where c is a constant expressing the ratio of the increase in radius of the electron at each pulse to its least radius (i.e. about 1/50) and the mass of ether extruded by 1 g of neutral matter being 2.9×10^{-4} g.

5.5. Source–sink theories of gravitation

There was another line of ether theories being developed at the same time as the pulsation theories. Whereas in the latter attraction and repulsion were produced by periodic flows of ether, the periods being of extremely short duration, this set of theories was based on secular flows of ether. In 1853 Riemann had suggested that a flow of ether through every particle into a greater universe could account for gravitation (cited by de Tunzelmann 1910, p. 443; see Riemann 1876, p. 503), but the first person to give a theoretical treatment of the idea was Karl Pearson in 1891. Before this Pearson had published a number of papers on the pulsatory theory.

The first of these, written in 1883, mainly dealt with practical applications such as the motion of ships (Pearson 1885). In this next paper, written in the same year and read to the Cambridge Philosophical Society in 1885, Pearson put forward a detailed gravitational theory (Pearson 1889). This, as we might have expected, gave a Clairaut hypothesis for the gravitational force, the extra term being in r^{-5}, but had certain novel properties. It appeared that the mass of a body changed slightly under the influence of other bodies in its neighbourhood owing to the alteration of periods of pulsation. For instance measuring the density of the Earth might produce different results when different test materials were used. Pearson thought that this unlikely consequence might necessitate a reinterpretation of his force law in terms of some other function than mass, but noted the occurrence of 'Baily's experiments [in which] the average mean density as obtained from a set of experiments with one kind of balls on the torsion rod differed from that obtained from experiments with balls of a different substance' (Pearson 1889, p. 115). The theory also resulted in a mass dependence on temperature, a dependence which was 'observed' for a time by P.E. Shaw in 1916 (see Chapter 6).

The second part of this paper was more powerful (Pearson 1888–9).† It was written against William Thomson, who in 1884 had proposed a mechanical model of the atom having 'a material core surrounded by a number of material spherical shells, linked together by elastic springs', and Lindemann who in 1888 had published a memoir giving the results of his investigations into Thomson's mechanical atom. Pearson's aim, and he thought that he had succeeded, was

> . . . to show that the results obtained by Lindemann are in no way dependent on the structure of the shell atom of Thomson; that, further, the pulsating spherical atom is better adapted than the shell atom to explain chemical association and disassociation, while, as shown in my first paper, it succeeds in throwing a certain amount of light on chemical, cohesive, and even gravitational force. (Pearson 1888–89, p. 60)

In 1891 Pearson abandoned the pulsatory interpretation in favour of the notion of 'ether squirts', although he wrote no more on these subjects (Pearson 1891). In 1870 William Thomson had discussed the forces between fluid sources and sinks and had noted the analogy with gravitation, but Pearson was putting this forward as a serious model:

> The ether squirt as a model dynamic system for the atom seems at any rate to possess the property of simplicity. But the action of one group of ether squirts upon a second group leads to equations the complexity of which seems quite capable of paralleling any intricacy of actual Nature . . . The law of gravitation and the theory of potential are shown to be more intelligible on the ether squirt theory than that of the pulsating sphere . . . . (Pearson 1891, p. 309)

As for the opponents and their shells and springs:

> The Thomson–Lindemann atoms and molecules thus show us so far only complex mechanisms, and raise the not unnatural repugnance of the philosophical mind to a dualistic theory of the universe. (Pearson 1891, p. 310)

His own theory was monistic:

> . . . the fluid medium in irrotational motion is the primary substance, the atom or element of matter is a squirt of the same substance. From whence the squirt comes into three-dimensional space it is impossible to say; the theory limits our possibility of knowledge of the physical universe to the existence of the squirt. It may be an argument for the existence of a space of higher dimensions than our own, but of that we can know nothing, and we are only concerned with the flow into our own medium, with the ether squirt which we propose to term 'matter'. (Pearson 1891, p. 312)

Pearson obtained an expression for the velocity of ether flow as a series expansion. This contained a constant term for gravitation, forced periodic terms for chemical affinity and cohesion, and other vibrational terms for optical and electrical phenomena. The gravitational potential is then given as

$$\sum \frac{4\pi\, \rho\, a_s^2 a_{s'}^2\, {}_0v_s\, {}_0v_{s'}}{\gamma_{ss'}} = \frac{1}{4\pi\rho} \sum \frac{q_s q_{s'}}{\gamma_{ss'}} \tag{5.12}$$

† In this paper Pearson referred to Thomson's Baltimore Lectures and the work of Lindemann (1888).

where ρ is the density of the medium, a_s the radius of sphere s, ${}_0v_s$ the mean velocity of flow of the ether across the surface of the sphere s, $\gamma_{ss'}$ the distance between s and s', and q_s the mean rate at which ether is poured in at the sth squirt, i.e. $q_s = 4\pi a_s^2 \cdot \rho \cdot {}_0v_s$. Pearson was then able to obtain the inverse square law by simply defining mass to be the 'mean rate at which ether is poured in at any squirt' (Pearson 1888–9, p. 314). The potential energy is actually kinetic energy of the ether. The squirt can be conceived of as either source or sink, which gives the by now familiar positive and negative masses.

The problem of where the ether goes to or comes from in the squirt theory was answered in an entirely physical way by Schott (1906). This suggestion arose from the problem of explaining the fineness of spectral lines, which had caused Jeans to postulate unknown forces between electrons and for which Rayleigh had provided an incomplete kinematic answer. The trouble was that for an atomic ring of electrons:

> . . . a strictly steady motion is impossible when we restrict ourselves to the case of an invariable electron and exclude all forces not of electromagnetic origin. . . . The difficulty can be overcome by supposing the electron to vary slowly in radius. (Schott 1906, p. 23)

As an expanding electron is an ether source this fits with the Pearson theory, and Schott obtained an inverse square law applicable to gravitation (with a time-dependent constant of gravitation). Schott himself did not think that his theory could be satisfactorily worked out but de Tunzelmann viewed it with favour. He thought in addition that the deformation of electrons with speed would lead to a velocity-dependent flow of ether which could possibly account for the planetary perihelion motions. Even Schott's theory leads to the requirement of a greater universe for the explanation of the secular expansion of the electrons (which could be taken as the expansion phase of a periodic change with long period); de Tunzelmann suggested that the expansions were caused by an overall decrease of ether pressure in the universe. Unless this decrease could be accounted for internally (such as in general relativity where the expansion of the universe is a result of the field equations), one had to have recourse to a pressure decrease at the boundary of the universe caused by some agent in the outer universe.

5.6. Le Sage's theory of gravitation

The last gravitational theory to be considered is that of Le Sage. It was put forward by Le Sage in his *Newtonian Lucretius* of 1782, and brought back into prominence by William Thomson in 1873 when he showed how to circumvent an argument against the theory also given by Maxwell and later used by Poincaré (Thomson 1873).† It provoked widely differing reactions. John

† Thomson includes extracts from the Lucrèce Newtonien which had appeared in the *Transactions of the Royal Berlin Academy* (1782). A translation introduced by S.P. Langley is given in Abbott (1898).

Herschel thought it 'too grotesque to need serious consideration' (quoted by Taylor 1876), but P.G. Tait (a follower of Thomson) remarked that it was 'the only plausible answer to this [problem of explaining gravity] which has yet been propounded' (quoted by Taylor 1876).

Le Sage's theory answered certain questions that were left unasked in the discussion earlier in this chapter of the three-dimensional argument for the inverse square law. That argument gave no idea of the mass dependence of the law. It also did not answer why a large body and a small body could have the same weight. If a body at a distance r from a centre of force is taken to a distance $2r$, one can easily see from the three-dimensional argument that the intensity of force it is subject to is decreased by a quarter. However, if the body remains at the same distance and its shape is altered so that it occupies a greater proportion of the surface area of the sphere of radius r, then its weight does not increase in the same proportion. Indeed its weight does not increase at all and this is contrary to the three-dimensional argument taken at its face value. Le Sage's theory resolves these problems, or at least appears to do so.

Le Sage conceived of matter as being built up of indivisible particles in the form of cages with bars of extremely small diameter. Space was continually traversed by gravific particles of extremely high velocities in all directions and rarely collided with each other. An isolated body in space would not be moved by these gravific corpuscles because it received an equal number of impulsions (from the corpuscles hitting the cage bars) on all sides. If another body was brought up towards this previously isolated body, the latter would be shielded to a certain extent by the former from the corpuscles approaching from that direction. The equilibrium of impulsions thus disturbed, the bodies would be pushed towards each other as if they were attracted:

> It is not necessary to be very skilful to deduce from these suppositions all the laws of gravity, both sublunary and universal (and consequently also those of Kepler, etc.) with all the accuracy with which observed phenomena have proved those laws. Those laws, therefore, are inevitable consequences of the supposed constitutions. (Thomson 1873, p. 323)

The mass of a body enters the law of attraction because the more massive a body is the more cages it contains, and hence the more bars, so that it suffers more collisions with the gravific corpuscles. Changing the shape of a body does not affect its weight because one does not change the number of cages in it.

The corpuscles of Le Sage were perfectly hard. For the theory to work they had to have a lower velocity after collision than before, which implied a gradual rundown of gravitational action, but Le Sage felt this to be no objection to his theory so long as it took place sufficiently slowly.† With

† That these suppositions imply a gradual diminution of gravity from age to age was carefully pointed out by Le Sage, and referred to as an objection to his theory. Thus he says, '. . . Donc, la durée de la gravité seroit finie aussi, et par conséquent la durée du monde.

Reponse. Concedo; mais pourvu que cet obstacle ne contribue pas à faire finir le monde plus promptement qu'il n'auroit fini sans lui, il doit être considéré comme nul.' (Thomson 1873, p. 328)

perfectly hard corpuscles there is no rebound after collision, so they would stick with the struck body. That bodies did not continually increase in mass could be explained by sufficiently light corpuscles, but Le Sage preferred to think of them slipping off the body after collision. Also with perfectly hard corpuscles the maximum energy is converted into heat. Maxwell (1875) objected that with the magnitudes required to account for gravitation the heat gained by any body would raise it to a white heat, or in Thomson's earlier words 'this elevation of temperature would be sufficient to melt and evaporate any solid, great or small, in a fraction of a second of time' (Thomson 1873, p. 329). Thomson replaced the ultimate particles by vortex atoms of perfectly elasticity. Since in perfectly elastic collisions no kinetic energy is lost the above objection no longer held, but then if the gravific atom had the same velocity as before gravitation would not occur. Thomson suggested that the translatory kinetic energy was indeed decreased on collision but that the corpuscles retained the residual kinetic energy in the form of vibrational energy. This vibrational energy would be converted back into translational energy, thus restoring the gravitational power of action to its former level, and Thomson cited in support of this a principle of Clausius which held that in a system of interacting elastic corpuscles the ratio of vibrational and rotational energy to translational energy was a constant. After a gravitational collision this ratio would have been disturbed and, in accordance with Clausius's principle, would be restored by inter-corpuscle collisions.

Le Sage's theory and variations of it were investigated by a number of scientists in the second half of the nineteenth century, including Isenkrahe, Bock, Rysanek, Jarolimek, and the Englishman S. Tolver Preston. Preston (1877) saw the gravific corpuscles as akin to particles of a gas so that they obeyed the kinetic theory of gases which ensured an equal distribution of corpuscles in all directions. Rysanek (1888) used this idea and its consequences for the motion of Neptune to show that the velocities of the particles had to be between 2.7×10^{12} and 5.4×10^{17} m s^{-1}. Preston did not use Thomson's Clausius argument to circumvent the Maxwell argument, but thought that by indefinitely increasing the number of corpuscles the energy each contained could be made indefinitely small so that the energy transferred to any body would not be sufficient to increase its temperature to a measurable extent (Preston 1895, p. 151). This argument is not convincing without a quantitative reassurance that the gravitative action would remain sufficient. Preston pointed out that on Thomson's theory the action of gravity would not extend beyond a mean distance equal to that distance required for the full translational energy of a corpuscle to be restored. This was in no way an objection, for the inverse square law had only been verified up to 30 AU (Sun–Neptune distance) and the resulting lack of attraction between stars would increase the stability of the universe. We have seen in Chapter 4 how Seeliger used Le Sage's theory to account for the stability of the universe in answer to the gravitational analogue of Olbers' paradox, but he used the possibility of gravitational absorption inherent in this theory rather than Preston's idea.

A novel objection to Le Sage's theory appeared in 1897 (Farr 1897). The theory required that absorption of gravific particles was very small, not only because absorption was not observed—Seeliger's amount did not have to be observable at planetary distances—but because beyond a certain bulk (at which all corpuscles were observed) attraction would become mass independent. Farr objected that owing to a result of Nernst the molecules of liquids at their boiling points and at atmospheric pressure occupied about 0.3 of the total volume, a proportion so great that an unacceptable amount of screening would occur. This seems a strong objection so long as molecules are taken as solid and opaque to gravific particles.

Darwin (1905) gave a mathematical treatment of Le Sage's theory. He obtained expressions for the forces due to both the normal and tangential components of the impacts which resulted in the following expression for the force between two spheres:

$$F = \frac{1}{8}\pi\rho\frac{v^2a^2b^2}{R^2}\left\{(k+k')+\frac{8}{3\times 5}\frac{a}{R}(k-k')+\frac{4a(4a+7b)}{3\times 5\times 7\,R^2}(k-k')\right.$$
$$+\frac{a(8a^4+36a^2b^2+21b^4)}{3\times 5\times 7\,R^5}(k-k')$$
$$\left.+\frac{a}{R^7}\left(\frac{32}{693}a^6+\frac{8}{21}a^4b^2+\frac{4}{7}a^2b^4+\frac{1}{6}b^2\right)(k-k')\right\} \qquad (5.13)$$

where ρ is the density of the medium, v is the velocity of the corpuscles, a and b are the radii of the spheres concerned, R is the distance between their centres, and k and k' are elasticity factors (k' is for the tangential component) such that $k=1$ for perfectly inelastic corpuscles and $k=2$ for perfectly elastic corpuscles. This is not a rigorous inverse square law but it may be made so if one puts $k=k'$. For the general law action and reaction are not equal since a and b do not enter it equally, but this could be avoided if these spheres are taken as elementary particles of matter and if all such elementary particles of matter are of equal size. The above force law is still approximate as Darwin neglected reflected particles. Although this law will give a perihelion motion, because the coefficient of the term in R^{-3} is of the order of the radius of an elementary particle the motion will be very small.

If Darwin's work were ever thought of as giving a certain air of respectability to Le Sage's theory, and Darwin himself did not take that view,†

† Darwin started his *Introduction to dynamical astronomy* (1916) with a review of some theories of gravitation. Referring to the Maxwell argument against Le Sage's theory, Darwin continued 'Lord Kelvin has, however, pointed out that there is a way out of this fundamental

then it was Poincaré (1908, and 1953, pp. 203–216) who finally laid it to rest. This he did by putting on a quantitative basis the old objection that gravitating bodies would receive too much heat. Poincaré related it to the resistance suffered by bodies using as a common factor the velocity of the corpuscles. Using a Laplace-type argument, Poincaré showed that this had to be greater than 24×10^{17} times the velocity of light, and that this led to the impossible consequence of the Earth receiving from the corpuscles 10^{20} times the amount of heat the Sun emitted in all directions. Poincaré found no better situation in considering the corpuscles as imperfectly hard, or considering a wave version of the theory in which radiation pressure (Maxwell–Bartholi pressure) replaced the corpuscular action, a version considered by Darwin and held by Tommasina. Le Sage's theory would not work:

> The fact that a hypothesis so artificial, and containing so many defects, as that of LeSage should be deemed worthy, after the lapse of more than two hundred years, of serious investigation by such eminent physicists, will serve to impress the reader's mind with the extreme difficulty presented by the problem of forming a dynamical representation of gravitational action. (de Tunzelmann 1910, p. 433)

So wrote de Tunzelmann on Le Sage's theory. Yet if all the theories we have considered were abject failures, it was the act of reflecting on the nature of gravitation itself, of probing behind it rather than just accepting the inverse square law or suitable alternatives, that led Einstein to general relativity. If they were failures they were at least being taken seriously, and I have considered them because it seemed that only by their help could one assess the *a priori* likelihood of Hall's hypothesis or gain other clues to the explanation of the motion of Mercury's perihelion. We have seen that on none of the theories was a hypothesis of the type that Hall suggested obtained, and I conclude from this and from the two previous arguments for the inverse square law that Hall's hypothesis is *a priori* unlikely.

5.7. Newcomb's rejection of Clairaut's hypotheses: 1882

The hypothesis that was apparently supported was that of Clairaut. From *a priori* grounds it seems that Newcomb would have made a more likely choice by using this instead of Hall's hypothesis in his work of 1895, but he had in fact considered it in his earlier paper on the perihelion of Mercury (Newcomb

difficulty . . . on the other hand gravitation will not be transmitted to infinity, but only to a limited distance. I will not refer further to this conception save to say that I believe that no man of science is disposed to accept it as affording the true road'.

Of Bjerknes' experiments, 'however curious and interesting these speculations and experiments may be, I do not think that they can afford a working hypothesis of gravitation'.

Of Osborne Reynolds' ideas, 'Notwithstanding that Reynolds is not a good exponent of his own views, his great achievements in science are such that the theory must demand the closest scrutiny' (Reynolds (1903, p. 205) obtained an inverse square law of attraction).

1882). Then Clairaut's hypothesis had been rejected using a very strong argument that had been foreseen by Clairaut himself:

> A term of the inverse third power which, at the distance of Mercury, should have a value even the millionth part of the total gravitative force of the sun would, at the distance of a foot, have a value two hundred thousand times that of the term depending on the inverse square. If higher powers than the cube were added the discrepancy would be yet more enormous. The existence of such a term of such magnitude is out of the question. (Newcomb 1882, p. 476)

Newcomb's figures are both correct and applicable. We know from Newton that a force of the form $bA^{m-3} - cA^{n-3}$ gives an angle between apsides of $180°\{(b-c)/(mb-nc)\}^{\frac{1}{2}}$. For Clairaut's hypothesis with an inverse third power term, we have $A^{-2}+aA^{-3}$; $b=1$, $a=-c$, $m=1$, and $n=0$. Then the angle between successive perihelia is $360°(1+a)^{\frac{1}{2}}$. The advance in the perihelion of Mercury is 40″ per century or 0″.4 per year or about 0″.1 per revolution. Therefore we require

$$360\,(1+a)^{\frac{1}{2}} = 360+0.1/3600$$

or

$$(1+a)^{\frac{1}{2}} = 360.000\,0278/360 \approx 1+8\times10^{-8}$$

so that

$$1+a = 1+16\times10^{-8} \text{ by squaring both sides,}$$

$$a = 16\times10^{-8}.$$

Newcomb's estimate is about right and the objection is fatal. In the Cavendish experiment one measures the gravitational force between two lead balls placed near to each other by observing the gentle twisting of the supporting torsion wires. With the addition of the inverse third power term at the magnitude required for the perihelion advance of Mercury the total force would be vastly greater than that producing the gentle twisting.

In 1895 Newcomb was therefore left with Hall's hypothesis as the only possible theory containing terms which were functions of distance alone. When the 1903 lunar theory of E.W. Brown showed that this hypothesis was untenable, Newcomb switched his support to Seeliger's hypothesis. The gravitational theories in contention at this time, apart from those of this chapter, were the velocity-dependent forces which are the subject of the next chapter, and the earliest of the relativistic theories which will be discussed in Chapter 7.

6

VELOCITY-DEPENDENT FORCE LAWS OF GRAVITATION

6.1. Velocity-dependent force laws

Velocity-dependent force laws of gravitation contain terms which depend on the velocities of the bodies concerned as well as their distance apart. The velocity terms can be seen as providing a small addition to the standard Newtonian force so that a perihelion advance is obtained even in the case of a single planet in orbit around the Sun. The velocity-dependent force laws played a more central role in the history of the anomalous advance in the perihelion of Mercury than the laws found in Chapter 5. They gained at least the attention of some working astronomers since each law predicted a substantial perihelion advance which gave some hope of explaining the whole Mercury anomaly. In addition they drew their inspiration from the increasingly successful science of electrodynamics. As electrodynamics progressed and produced new force laws, their gravitational equivalents were investigated. The more successful electrodynamics became, the more legitimate the new gravitational laws appeared.

The school of action-at-a-distance electrodynamics took the action between distant electric particles to be fundamental and expressed this action in terms of a force law. The action came to be interpreted as being propagated with the velocity of light, and in the ballistic theories fictitious particles were invoked to transmit the action. These particles travelled with a velocity equal to the sum of the velocities of their source and light, and so came into conflict with special relativity which postulated that the velocity of light (electromagnetic action) was constant and independent of the velocity of its source. The school of action-at-a-distance electrodynamics had started with Wilhelm Weber who had published his fundamental force law in 1846. The law was followed by others from Gauss, Riemann, and Clausius, and each was translated into a gravitational form. Even the rival approach, that of the field theories of Maxwell, was investigated for gravitational use. At the turn of the century Ritz produced a sophisticated action-at-a-distance force law. He gave its gravitational equivalent and showed that it could give the full perihelion advance for Mercury. Ritz soon outlined a much more ambitious theory which attempted to cover a comprehensive range of phenomena, but he died soon after its publication and few scientists showed interest in the theory.

Astronomical arguments were given which purported to refute it on the grounds that it contained a variable velocity of light. Another action-at-a-distance theory of gravitation had been published by this time. Gerber's theory provoked some reaction when it was shown that it produced an identical result to general relativity for the perihelion advance of a planet, but it was never a serious competitor.

The roots of all these laws lay back in the nineteenth century with the work of Wilhelm Weber. Because the contact between the electrodynamic and gravitational force laws was so strong it is important to outline the development of the electrodynamics before the gravitational applications are treated.

6.2. The electrodynamical background

Wilhelm Weber was for most of his working life a professor of physics at Göttingen, where Gauss also lived and worked. The two were friends and colleagues—for instance in 1833 they built together one of the first electric telegraphs. For political reasons Weber was forced to leave Göttingen in 1833, being re-instated in 1849, and from 1843 until 1849 he occupied a chair at Leipzig. While at Leipzig, in 1846, he published his force formula which we know as Weber's law (Weber 1846, 1848).

The starting point for the derivation of Weber's law was Ampère's law of 1825 for the force between two current-carrying circuit elements. The force $\mathbf{dF}$ exerted by the circuit element $\mathbf{dS}$ (current strength i) on the circuit element $\mathbf{dS}'$ (current strength i', relative position $\mathbf{r}$) is given by

$$\mathrm{d}\mathbf{F} = \frac{ii'}{r^2}\left\{3(\mathrm{d}\mathbf{S}\cdot\hat{\mathbf{r}})(\mathrm{d}\mathbf{S}'\cdot\hat{\mathbf{r}}) - 2(\mathrm{d}\mathbf{S}\cdot\mathrm{d}\mathbf{S}')\right\}. \tag{6.1}$$

Weber considered the current as consisting of an equal quantity of negative and positive particles of electricity moving in opposite directions with equal speeds, such that particles of the same sign attracted if they moved in the same direction and repelled if they moved in opposite directions. This hypothesis was due to Fechner, also of Leipzig, in 1845. In this way Weber introduced particles into Ampère's formula which, after some substitution and manipulation, split into four similar sets of terms corresponding to the forces of each current stream on the two current streams in the other circuit element. These were used to infer that the force between two electric masses E and E' at a distance R apart was given by

$$F = \frac{EE'}{R^2}\left\{1 - \left(\frac{\mathrm{d}R}{\mathrm{d}t}\right)^2 + 2R\frac{\mathrm{d}^2R}{\mathrm{d}t^2}\right\}. \tag{6.2}$$

where the first term, representing the electrostatic attraction as expressed in Coulomb's law, has been added because it is not found in Ampère's law which is for circuit elements of neutral charge. Weber was able to deduce from his force law an expression for the electromotor force found in voltaic induction. Since this involved variable currents, whereas Ampère's law contained steady currents, the success of this deduction was confirmation of the validity of the generalization of Coulomb's law to the case of moving electric charges. In a consistent system of units Weber's law is given by

$$F = \frac{ee'}{r^2}\left\{1+\frac{r}{c^2}\frac{\mathrm{d}^2r}{\mathrm{d}t^2}-\frac{1}{2c^2}\left(\frac{\mathrm{d}r}{\mathrm{d}t}\right)^2\right\}. \tag{6.3}$$

Measurements made by Weber and Kohlrausch in 1856 showed that c, a constant with the dimensions of a velocity, had almost the same value as the velocity of light.

In his *Treatise* Maxwell gave a derivation of Weber's law from Ampère's law, and showed that Fechner's hypothesis was sufficient but not necessary for Weber's law (Maxwell 1873, p. 482). Later Clausius (1877) disputed this and showed that if a one-current hypothesis were used, i.e. a hypothesis in which one type of charge moved while the other remained at rest, Weber's law would imply that a current-carrying wire exerted a force on a charge at rest. Clausius was supported by Lorberg (1878), but we can see that the objection is not fatal for the force is proportional to $(w/c)^2$ where w is the velocity of the current. We now know that w is of the order of a few millimetres per second, so that the force would be extremely small, but the current velocity could not be measured until 1879 when the Hall effect was discovered. In 1880 A. von Ettingshausen found a value of 0.1 cm s^{-1}, which is sufficiently small to allow the Clausius objection to be overcome. Clausius put forward his own force law, which will be discussed below, according to which a current-carrying wire exerted no force on a stationary charge.

Weber's law was strongly criticized by Helmholtz, whose initial objection was that it violated the principle of the conservation of energy (or force as it was known then). The root of this lay in his paper of 1847, *On the conservation of force*, in which he had purported to prove that

> if . . . natural bodies are possessed of forces which depend on time and velocity, or act in other directions than the lines which unite each two separate material points, for example, rotatory forces, then combinations of such bodies would be possible in which force might be lost or gained *ad infinitum*. (Helmholtz 1847, p. 131)

However, Helmholtz was considering that potentials were functions of position alone, which they were held to be before Weber. Weber extended the notion of potential to include a velocity dependence and had in fact already published the required potential for his force law in the original paper proposing the law, giving it as

$$\psi = \frac{ee'}{r}\left\{1-\frac{1}{2c^2}\left(\frac{\partial r}{\partial t}\right)^2\right\} \tag{6.4}$$

From this we obtain the force

$$F = -\frac{\partial \psi}{\partial r} = \frac{ee'}{r}\left\{1 - \frac{1}{2c^2}\left(\frac{\partial r}{\partial t}\right)^2 + \frac{r}{c^2}\frac{\partial^2 r}{\partial t^2}\right\}.$$

Weber's law does then conserve energy, but Helmholtz's criticism was at first accepted by Maxwell. In this first major paper on electricity, *On Faraday's lines of force*, which was published in 1856, Maxwell wrote concerning Weber's theory:

There are also objections to making any ultimate forces in nature depend on the velocity of the bodies between which they act. If the forces in nature are to be reduced to forces acting between particles, the principle of the Conservation of Force requires that these forces should be in the line joining the particles and functions of the distance only. (Maxwell 1890, p. 208)

This was despite the publication of Weber's potential in 1846, and the publication of an English translation of the 1848 paper containing the essentials of the 1846 paper in Taylor's *Scientific Memoirs* in 1852. In the introduction to his *A dynamical theory of the electromagnetic field* of 1864 Maxwell still objected to the Weber approach:

The mechanical difficulties, however, which are involved in the assumption of particles acting at a distance with forces which depend on their velocities are such as to prevent me from considering this theory as an ultimate one, though it may have been, and may yet be useful in leading to the coordination of phenomena. (Maxwell 1890, p. 527)

In his *Treatise* of 1873, Maxwell recognized that the conservation of energy objection was invalid (Maxwell 1873, p. 484). By this time Helmholtz had given two related objections, and Weber had begun to defend his theory in print. The first of Helmholtz's new arguments appeared in 1870 (Helmholtz 1870, p. 67). One can write Weber's law as

$$m\frac{\mathrm{d}^2 r}{\mathrm{d}t^2} = \frac{ee'}{r^2}\left\{1 - \frac{1}{c^2}\left(\frac{\mathrm{d}r}{\mathrm{d}t}\right)^2 + \frac{2r}{c^2}\frac{\mathrm{d}^2 r}{\mathrm{d}t^2}\right\}.$$

Multiplying by $\mathrm{d}r/\mathrm{d}t$ and integrating gives

$$\frac{m}{2}\left(\frac{\mathrm{d}r}{\mathrm{d}t}\right)^2 = C - \frac{ee'}{r} + \frac{ee'}{rc^2}\left(\frac{\mathrm{d}r}{\mathrm{d}t}\right)^2$$

or

$$\frac{1}{c^2}\left(\frac{\mathrm{d}r}{\mathrm{d}t}\right)^2 = \frac{C - ee'/r}{\frac{1}{2}mc^2 - ee'/r} \tag{6.5}$$

Helholtz considered the conditions $ee'/r > \frac{1}{2}mc^2 > C$, where C is the constant of integration. In this case the right-hand side of equation (6.5) is greater than unity, so that $\mathrm{d}r/\mathrm{d}t$ is real. If this is positive, then 'r will increase until $ee'/r = \frac{1}{2}mc^2$, when $\mathrm{d}r/\mathrm{d}t$ becomes infinitely large'. Weber (1871, pp. 296–9) replied by considering what this situation involved. First of all if the right-hand side of equation (6.5) were greater than unity, then $\mathrm{d}r/\mathrm{d}t > c$, where

c is for the form of the law given here, equal to $\sqrt{2}$ times the velocity of light. Such a velocity, Weber pointed out, was not to be found in nature. Secondly the condition $ee'/r > \frac{1}{2}mc^2$ required r (the distance between the two particles e and e') to be extremely small; indeed we can now see that for two electrons the maximum distance r given by $ee'/r = \frac{1}{2}mc^2$ is twice the classical radius of the electron, and the electrons could not start at a distance apart less than twice their radius. Weber finally pointed out that the same objection—that since $\mathrm{d}r/\mathrm{d}t \to \infty$ infinite kinetic energies could be produced in finite distances—could be made against ordinary gravitational theory when the gravitating masses were considered as point particles (Weber 1871, p. 298).

Helmholtz produced a second objection in 1872 (Helholtz 1872). He considered a fixed non-conducting spherical surface of radius a charged uniformly with a surface density σ. Inside, a particle of charge e and mass m moved with a velocity v. The electrodynamic potential of the particle can be obtained from equation (6.4) to give

$$\psi = 4\pi a\sigma e(1 - V^2/6c^2)$$

The equation of constant energy is then given by kinetic energy + potential energy = constant

or
$$\tfrac{1}{2}mv^2 + (\psi + V) \rightleftharpoons \text{constant}$$

or
$$\tfrac{1}{2}(m - \tfrac{4}{3}\pi a\sigma e/c^2)\mathrm{v}^2 + 4\pi a\sigma \mathrm{e} + V = \text{constant} \tag{6.6}$$

where V is the mechanical potential energy. Helmholtz pointed out that one could under certain conditions make the coefficient of v^2 in equation (6.6) negative, giving the particle a negative mass. Weber (1875, p. 334) replied that though this was possible, the physical circumstances were unlikely to occur. Sensible values of charge density required the radius of the sphere to be 3 million times the distance between the Earth and Sun!

By this time Weber's law had been joined by others, and a strong continental action-at-a-distance school was formed. The first 'new' law had actually been produced before that of Weber by his old friend Gauss who had obtained the following law in 1835:

$$F = \frac{ee'}{r^2}\left[1 + \frac{1}{c^2}\left\{u^2 - \frac{3}{2}\left(\frac{\mathrm{d}r}{\mathrm{d}t}\right)^2\right\}\right]. \tag{6.7}$$

In equation (6.7) F is the repulsive force between two particles of charge e and e' moving at a relative velocity u with a distance r between them where c is a constant with the dimensions of velocity which Weber and Kohlrausch in 1856 had found to be nearly the same as the velocity of light. The similarity of Gauss's law to that of Weber can be seen in Maxwell's *Treatise* (Maxwell 1873, Part 4, Chap. 23) in which he derived both from Ampère's law. The difference between them is that whereas Gauss's law contains a relative velocity, that of Weber contains time derivatives of a scalar distance. The difference is apparent when one considers circular motion; a particle moving in a circle about a fixed point has a relative velocity with respect to the centre, but because the distance between the centre and the particle is constant the time derivatives of the distance vanish. Whereas Weber's law does conserve energy with the potential Weber gave in 1846, Gauss's law does not do so.

It was published in 1867, after Gauss's death (Gauss 1867, pp. 616–7). He had been worried about being unable to form a clear representation of how the electrodynamic action passed from one particle to another (Gauss 1867, p. 629). Gauss conceived of this action-at-a-distance propagating with the speed of light, but Weber succeeded in deriving his own force law without any such notion, and used just Ampère's law and Fechner's current hypothesis. As we shall see below the action-at-a-distance school can be interpreted in terms of a finite propagation time, and the constant c (having the dimensions of velocity) which had entered Weber's law as a conversion factor between electrostatic and electrodynamic units is closely related to the velocity of propagation.

A third law was put forward by Riemann in lectures in 1861 and published in an edition of the lectures which appeared in 1875 (Hattendorf 1876). The force F was given by

$$F = \frac{ee'}{r^2}\left(1+\frac{u^2}{c^2}-2\frac{u}{c^2}\frac{\mathrm{d}r}{\mathrm{d}t}+\frac{2}{c^2}f\right) \tag{6.8}$$

where f is the relative acceleration between the two charged particles and the rest of the symbols are as before. Riemann's force is not directed along the line joining the two particles. It conserves energy with the following Lagrangian;

$$L_{\mathrm{R}} = \frac{-ee'}{r}\left\{1-\left(\frac{v}{c}\right)^2\right\}^{-\frac{1}{2}}. \tag{6.9}$$

This can be compared with Weber's Lagrangian

$$L_{\mathrm{w}} = -\frac{ee'}{r}\left\{1+\frac{1}{2c^2}\left(\frac{\partial r}{\partial t}\right)^2\right\}. \tag{6.10}$$

If one expands Riemann's Lagrangian to the second power of v/c, one obtains

$$L_{\mathrm{R}} = \frac{-ee'}{r}\left(1+\frac{v^2}{2c^2}\right)$$

so that the difference between the two is the difference between v and $\mathrm{d}r/\mathrm{d}t$—namely that $\mathrm{d}r/\mathrm{d}t$ is the component of v along the radius. Unlike Gauss's law, that of Riemann contains acceleration terms and hence can explain inductive effects. Like Weber's law it yields Ampère's law and also suffers from the Helmholtz negative mass objection.

The fourth law is that of Clausius, who as we have seen criticized Weber's law in 1877 for predicting a force on a stationary charge due to a current-carrying wire. The law he proposed was

$$F_x = -ee'\left[\frac{\mathrm{d}}{\mathrm{d}x}\left\{\left(1-\frac{vv'\cos\varepsilon}{c^2}\right)\frac{1}{r}\right\}-\frac{1}{c^2}\frac{\mathrm{d}}{\mathrm{d}t}\left(\frac{1}{r}\frac{\mathrm{d}x'}{\mathrm{d}t}\right)\right] \tag{6.11}$$

with similar expressions for the other two force components (Clausius 1877). The velocities v and v' here are absolute velocities, i.e. measured relative to an ether, whereas the three previous laws contained relative velocities. For Clausius this had the advantage that two charges moving at the same velocity relative to the ether, with a velocity relative to each other of zero, would experience a force in addition to the electrostatic force on his law but not on the others, and one would expect such a force because the two moving charges each constitute a current. Another difference is that the Clausius forces between two bodies are not equal and opposite so that the momentum of the system is not constant. In order to conserve momentum one must take into account the ether surrounding the bodies, assign a finite density to it and allow it to carry momentum. Clausius's law conserves energy and does not suffer from the Helmholtz negative mass objection.

By this time the action-at-a-distance school on the continent was being rivalled by the proponents of Maxwell's theory. Maxwell developed his electromagnetic theory in three main papers published in 1856, 1861–2, and 1864. The famous experiments of Hertz in 1888 seemed to confirm the primacy of the field concept as found in Maxwell's theory, which has become the apparent victor over the action-at-a-distance theories. However, this is not so, and the electron theory which we use today is due to Lorentz who in 1892 produced a synthesis of the opposing lines of thought. From the Clausius force law and using retarded potentials to express Maxwell's electric and magnetic fields **E** and **H**, he deduced the following expression for the force **F** on a charge e moving with a velocity **v** in fields **E** and **H**:

$$\mathbf{F} = e\mathbf{E}+\frac{e}{c}(\mathbf{v}\times\mathbf{H}). \tag{6.12}$$

The particle–particle interaction (action at a distance) was replaced by a particle–field field–particle interaction. One could question the need for the fields, and it has been held that we ought to use the pure form of the action-at-a-distance theories in their subsequent form. The crucial new concept was the retarded potential—a potential propagated with a finite velocity—which fulfilled the ideas of Gauss as found in correspondence to Weber and introduced by Riemann and Lorenz. One can deduce Maxwell's equations by using a retarded scalar potential and a retarded vector potential and definitions of **E** and **H** in terms of them. The retarded potentials were generalized by Liénard in 1898 and by Wiechert in 1900 to cover the case of moving particles, and Liénard gave the resulting expressions for **E** and **H**. In 1903 Schwarzschild obtained the force formula for two particles corresponding to Liénard's results. In 1908 Ritz gave an approximation to this formula in which terms of order greater than $1/c^2$ were neglected. The force formula one finally obtains is

$$\mathbf{F} = \frac{e_A e_B}{r^2}\left[\frac{\mathbf{r}}{r} - \frac{r^2}{2c^2}\left\{\frac{\dot{\mathbf{v}}_B}{r} + \left(\frac{\dot{\mathbf{v}}_B \cdot \mathbf{r}}{r^3} - \frac{\mathbf{v}_B{}^2}{r^3} + \frac{3(\mathbf{v}_B \cdot r)^2}{r^5}\right)\mathbf{r}\right\} + \frac{1}{c^2}\left[\frac{\mathbf{v}_A \cdot \mathbf{r} \times \mathbf{v}_B}{r} - \frac{\mathbf{r} \times \mathbf{v}_A \cdot \mathbf{v}_B}{r}\right\}\right] \qquad (6.13)$$

F is not central, and action and reaction are not equal†.

6.3. The gravitational application of the laws of Weber *et al.*

We have found five force laws—those of Gauss, Weber, Riemann, Clausius, and Ritz. Each of these was at some stage investigated as a possible replacement for Newton's inverse square law in celestial mechanics, and bearing in mind that non-inverse square laws give perihelion motions this line of work will now be traced.

The first of the laws to be applied to the planetary motions was that of Weber. In 1864 Seegers produced a dissertation at Göttingen with the title *On the motion and perturbations of the planets circling the Sun according to the electrodynamic law of Weber* (Seegers 1864). It is likely that this dissertation was prepared under the supervision of Weber himself. As far as the planetary motions were concerned, Weber had thought that the measurement of the constant c made by him and Kohlrausch had shown that no observable difference would result if his law were adopted as the gravitational law:

> There results furthermore from the large value of the constant c that has been found, the interesting consequence that such a dynamical part could also be added to the gravitational force of ponderable bodies . . . without this dynamical part of the force exhibiting the least perceptible influence on the motions of the heavenly bodies. (Weber and Kohlrausch 1857)

Seegers integrated the equations of motion using elliptical integrals and obtained an advance of the perihelion, but no application of this was made and no indication was given of the possible relevance of this result to the anomalous advance in Mercury's perihelion discovered by Le Verrier five years before in 1859.

Weber's law was the subject of another mathematical exercise by Holzmüller (1870), who later became the director of a Technical High School in Westphalia. Holzmüller used the Hamilton–Jacobi method, referred to the dissertation of Seegers, and like him obtained a perihelion advance without making any applications of it.

By this time Weber's law had been applied to the case of Mercury, though

† For a description of Ritz's work see the papers cited later and Hovgaard (1931–2) and O'Rahilly (1965).

the result was not published until 1872 when it appeared in Zöllner's *Ueber die Natur der Cometen*. It had been obtained by Wilhelm Scheibner, Professor of Mathematics at the University of Leipzig since 1868, who had worked with P.A. Hansen and had a long list of papers on perturbation theory to his credit by the time he died in 1908. In 1897 Scheibner published a paper on Weber's law in which he repeated the application to Mercury and cited his previous work (Scheibner 1897). According to this he had given a lecture to the *Gesellschaft der Wissenschaften* at Leipzig on February 6, 1869, entitled *Über die allgemeine Gültigkeit des Weber'schen Gesetzes*. Using the Weber–Kohlrausch value of the constant c (439 450 km s^{-1}), Scheibner's result for the advance in Mercury's perihelion was cited by Zöllner as being 6″.73 per century. In 1897 Scheibner corrected this to 6″.4868, and also found 7″.1302 using a value of $c = 421\,450$ km s^{-1}. Comparing Scheibner's value with that of the anomaly found by Le Verrier, Zöllner proposed in 1872 that Weber's law be adopted for gravitation:

> Since however, as shown, the Newtonian law is contained as a special case in Weber's law, and the objections concerning the permissibility of this law from the standpoint of the principle of conservation of energy are removed, Weber's law must be assumed in place of the Newtonian law for the reciprocal action of stationary and moving material particles according to the rules of rational induction. (Zöllner 1872)

A few years later Zöllner conceived a theory in which the use of Weber's law found a natural place. He found that Mossotti had put forward a similar theory in 1836, and the complete exposition of the Zöllner–Mossotti–Weber theory contained a reprint of Mossotti's original paper as well as contributions from Weber (Zöllner 1882). The idea of Zöllner was that the interaction between two positively charged particles was slightly different from the interaction between two negatively charged particles. The electric interaction between two neutral bodies would not on this theory be zero, and there would remain a residual force between them of much smaller magnitude than the electric forces. This residual force, it was proposed, could account for gravitation. Weber showed that if the electric interactions were expressed by Weber's electrodynamic law the residual 'gravitational' force would also be expressed by a Weber-type law. This was an inverse square law with two small additional terms representing the velocity dependence and resulting in the 7″ per century perihelion advance for Mercury. Weber's law found a natural place in this theory in the sense that it was an electrodynamic law–any of its rivals could have been so used, and when Lorentz investigated the theory in 1900 he put it into a Maxwellian form.

Meanwhile Weber's law had been discussed on a much more phenomenological level. The French mathematical astronomer Tisserand (1872) investigated the use of this law and its effect on the planetary motions. He contrasted the rigorous integration carried out by Seegers with his own method which consisted of taking the dynamical part of Weber's law as a perturbing force added to the Newtonian inverse square force and applying classical perturbation theory. Tisserand found that periodic variations were obtained for the perihelion, eccentricity, and longitude of the epoch, but that these were of small amplitude (less than 0″.003 for Mercury). The only secular variation was obtained for the perihelion, and this was given by

$$\delta\varpi = \frac{f\mu}{h^2}\frac{n}{a}t \tag{6.14}$$

where $\delta\varpi$ is the advance in the perihelion in time t, f is the constant of universal gravitation, μ is the sum of the masses of the planet and the Sun, h is the velocity of propagation of the gravitational action, n is the planet's mean motion and a its mean distance. Using the Weber–Kohlrausch value of $h = 439\,450 \times 10^6$ mm s^{-1} Tisserand calculated the centennial advance in Mercury's perihelion to be 6″.28 and that in Venus's perihelion to be 1″32. Taking h to be the velocity of light he obtained 13″.65 and 2″.86 respectively. The form of Weber's law that Tisserand had used is the following:

$$F = \frac{fm\mu}{r^2}\left[1 - \frac{1}{h^2}\left(\frac{\mathrm{d}r}{\mathrm{d}t}\right)^2 + \frac{2}{h^2}r\frac{\mathrm{d}^2r}{\mathrm{d}t^2}\right\}. \tag{6.15}$$

If we compare equation (6.15) with the form given in equation (6.3), we find the two be equivalent if $2/h^2 = 1/c^2$ or $h = \sqrt{2}\,c$. Now c is the velocity of light or at least nearly so; Weber and Kohlrausch had measured h, or $\sqrt{2}\,c$, and not c itself, and their value of h was very nearly the product of $\sqrt{2}$ and the velocity of light. Because the value for the constant for the electrodynamic case had been directly measured by Weber and Kohlrausch we cannot allow any latitude to be taken in using different values. The constant is a conversion factor from electrostatic to electrodynamic units, but with the development of the electromagnetic theory in which electromagnetic actions are propagated with the velocity of light the earlier ideas of Gauss would have been somewhat vindicated. Propagated action at a distance lay beneath Weber's work, and Seegers (1864) took h to be the velocity of propagation. However, if one takes the gravitational case to be the same as the electrodynamic case, then the appropriate form of the law must be used depending on whether one uses c or $\sqrt{2}\,c$ as the velocity constant. If one uses c, then equation (6.3) must be used, and if one uses $\sqrt{2}\,c$ then equation (6.15) is appropriate. Weber's law properly gives 6″.28 and not 13″.65 for Mercury's perihelion advance, and Tisserand's paper was the source of much error on this point.

Interpreting the velocity constant purely as a conversion constant between static and dynamic units, one could have held that there was no reason why the gravitational constant should have the same value as in the electric case. The two types of force are sufficiently different (in magnitude, absence of negative masses, etc.) for this to have some plausibility. Interpreting it as a propagation factor gives this less plausibility, especially if a propagation through the ether is envisaged, since two different velocities require two ethers. Different values of the constant would also destroy the proposed unity between gravity and electrodynamics.

Propagated gravitational action had already been considered by Laplace (1805) in the *Mécanique céleste* where he had shown that aberration effects would be expected. He used a model in which attraction was caused by a gravitational wave flowing into the attracting body; applying this to the Moon's motion about the Earth he found a secular acceleration of the Moon's mean motion. This had in fact been observed, but Laplace himself had shown that it could be completely accounted for by ordinary Newtonian theory so that the velocity of the gravitational action had to be greater than 100 million times the velocity of light. However, in the 1850s the English astronomer John Couch Adams had found that Laplace's analysis was wanting and that there was indeed a residual 12″ in the secular acceleration. At this stage no one claimed that this was evidence for a velocity of gravitational action equal to a million times the velocity of light; instead it was attributed to tidal friction, despite the fact that this was a qualitative attribution not supported by quantitative work until the 1920s.

Lehmann-Filhés (1895) did investigate the lunar motion and showed that the acceleration could be accounted for by a velocity of propagation equal to a million times that of light, but he hardly pressed for this adoption. Of course, since the tidal friction explanation was generally accepted but was not yet quantitative, one could not be sure what the actual residual secular acceleration was. Lehmann-Filhés (1885) had considered planetary motions and the effect on them of propagated action. Though he cited the 'still unexplained' motion of Mercury's perihelion and showed that one could obtain only a perihelion motion, he neither gave qualitative results nor urged the adoption of propagated action. The ideas of Lehmann-Filhés, and also those of von Hepperger (1888), fall under the scope of Laplace's argument, but this is not necessarily so for all theories using propagation since the argument is only valid on the simple model of propagation assumed. When Poincaré discussed Lorentz's modification of the Zöllner–Mossotti–Weber theory, he pointed out that Lorentz's hypothesis was entirely different from that assumed by Laplace, and that 'the result of Laplace proves nothing against the theory of Lorentz' (Poincaré 1908; see Poincaré 1954, p. 579). When he discussed his own Lorentz invariant gravitational theory, he favoured a version in which the velocity of propagation was that of light.

> It seems at first that this hypothesis must be rejected without examination. Laplace has shown, in effect, that propagation is either quite instantaneous or much more rapid than that of light. But Laplace had examined the hypothesis of the finite velocity of propagation *ceteris non mutatis*; here, on the contrary, this hypothesis is complicated by many others, and it may happen that there was among them a more or less perfect compensation, as those of which the applications of the Lorentz transformation have already given us plenty of examples. (Poincaré 1906; see Poincaré 1954, p. 544)

The use of Weber's law does not then suffer from Laplace's argument. What it did suffer from was the fact that it gave only about one-sixth of the anomalous advance of Mercury's perihelion. Tisserand had not mentioned this anomaly and had not discussed the merits of adopting the law in celestial mechanics. The American astronomer C.H.F. Peters did do so. In 1879 he published a paper attacking the supposed observations of Vulcan (Le Verrier's infra-Mercurial planet) that had been made by Watson and Swift (Peters 1879). Peters showed that their observations had probably been of stars Theta and Zeta Cancri, and that none of the Vulcan observations carried any weight. The existence of a large number of invisible planetoids of the size of meteors had been suggested on the basis of the 39″ advance, but Peters claimed that even this was not a sufficiently sure foundation on which to rest such an assertion. He thought that 20″ could suffice since the advance had largely been determined by old transit observations of limited accuracy:

> Even if a motion of 20″ should be proved to certainty, this is no more than yet may be accounted for by various other causes, without resorting to a mass inside, and moving nearly in the place of Mercury,—an additional, somewhat forced supposition, required since node and mean motion are not affected.—The value of the mass of Venus, employed by Le Verrier, is not as certain as he presumed. Lately (*Astr. Nachr.* Nr. 2105) Dr. Powalky found it, from a discussion of the fine series of meridian circle observations of the Sun made at Dorpat, 1/25th larger, which substituted would bring alone 15″ increase in the motion of Mercury's perihelion.—The whole theory of Mercury also ought to be worked out with regard to the terms, with which W. Weber has supplemented the Newtonian law of gravitation for bodies when in motion, and which become sensible in the case of a fast moving planet like Mercury. Prof. Scheibner has computed that they produce a change in the perihelion of Mercury of 6.7″ per century, if the numerical constant be taken the same as Weber's electrodynamical experiments gave it (439 000 kilometres in a second). (Peters 1879, cols. 339–40)

But however good a single set of observations, it has to be used in conjunction with other sets, and Powalky's mass was never adopted. Thus there still remained a substantial perihelion motion that was anomalous. Peters's optimism was not shared by Simon Newcomb, who in 1882 discussed the transits of Mercury which had occurred between 1677 and 1881 (Newcomb 1882). The discussion therefore included more transits than had been available to Le Verrier, but was not as extensive as his important work of 1895. Newcomb found in 1882 that Le Verrier's value for the anomaly should be increased slightly to 42″.95. In his *Speculation on possible causes of the excess of motion of the perihelion of Mercury*, Newcomb gave Tisserand's results on the use of Weber's law. They were merely stated, and then followed by the sentence 'Objections have been raised to Weber's whole theory on the part of physicists, to whom the discussion of its possibility must be left' (Newcomb 1882, p. 477).

We have seen that there were objections to Weber's theory. It had been held not to conserve energy, it had been subject to Helmholtz's attacks of 1870 and

1872, it had by 1882 the rival laws of Riemann and Clausius, and Maxwell's theory had also appeared. Although the first objections had been shown to be either false or weak the existence of rivals may have deterred Newcomb from taking one in preference to the others, though he showed no sign of investigating the other laws. He may have felt uneasy about the velocity dependence of Weber's potential, though by 1873 the Göttingen mathematician Schering had used such potentials in mechanics (see Goldstein 1950).

Even if Newcomb had felt no objection to Weber's law, there still remained 35″ of the perihelion advance to account for. Newcomb had shown that the 20″ of Peters was half the amount needed to satisfy the transit observations. The only alternative to changing the inverse square law was to postulate disturbing matter, but anyone who objected to postulating sufficient to cause an advance of 42″ on the grounds of invisibility would also object to postulating sufficient to produce 35″. Weber's law seemed not to give a large enough advance, and it now had to be asked whether the other laws produced a better result. This question was in fact answered in 1890 and 1895.

Tisserand (1890) returned to this problem and calculated the advance caused by Gauss's law. One might wonder why he chose 1890, which was thirteen years after the hypothesis of a single intra-Mercurial planet died along with Le Verrier and eight years after Newcomb had confirmed the reality of Mercury's perihelion advance. It is likely that he came across Gauss's law in Bertrand's *Sur la théorie mathematique d'électricité* (1890) which Tisserand referred to in this paper as 'le bel Ouvrage de M. Bertrand'. Tisserand repeated his work on Weber's law, giving the value of 14″.4 for the advance of Mercury's perihelion. This should have been 7″.2 for the reason already pointed out. He committed the same error with Gauss's law, claiming it gave an advance of 28″.2 for Mercury instead of the actual 14″.1 obtained using the velocity of light as the velocity of propagation which Tisserand intended to do. This had the effect of making Gauss's law appear more successful that it actually was, leaving only about 14″ to be accounted for by disturbing matter. Tisserand noted that the single Vulcan hypothesis had had to be abandoned, and that this left the possibility of an asteroid ring:

> The question remains open without being resolved. It is curious to notice that the law of Gauss will explain 3/4 of the excess in question (it will explain it completely if one gave a value equal to 6/7 of the velocity of light), without moreover troubling in an appreciable way the agreement realised by the law of Newton in the theory of celestial motions. I confine myself to pointing out this coincidence without claiming at all that Gauss's law is the truth. (Tisserand 1890, p. 314)

This is just as well, for Gauss's law gave three-eighths of the excess and not three-quarters, and also did not conserve energy. He would not have realized this though if he had read Bertrand's book, for it is claimed there that Gauss's law is identical with that of Weber, the latter being deducible from the former (Bertrand 1890, p. 183) However, Tisserand would have realized that different consequences were found for each by virtue of his previous calculations, so that they could not have been identical.

In this same year Lévy (1890a) investigated the possibilities of combining the laws of Weber and Riemann. Denoting the respective potentials by P_1 and P_2 and noting that both gave the same result for two closed electric currents, he realized that the difference $P_2 - P_1$ would, if it was the actual potential, give no effect. He therefore proposed the potential

$$P = P_1 + \alpha(P_2 - P_1) \tag{6.16}$$

where α is some as yet undetermined constant. This potential gives exactly the same result as Weber's potential P_1 and hence also as Riemann's potential for the observed phenomena of electric currents. The combined potential investigated by Lévy is then

$$P = \frac{fm\mu}{r}\left\{1 - \frac{1}{h^2}\left(\frac{\mathrm{d}r}{\mathrm{d}t}\right)^2\right\} + \alpha\frac{fm\mu}{r}\left[-\left\{1 - \frac{1}{h^2}\left(\frac{\mathrm{d}r}{\mathrm{d}t}\right)^2\right\} + \left\{1 - \frac{1}{h^2}v^2\right\}\right] \tag{6.17}$$

$$= \frac{fm\mu}{r}\left\{1 - \frac{1}{h^2}\left(\frac{\mathrm{d}r}{\mathrm{d}t}\right)^2\right\} + \alpha\frac{fm\mu}{r}\left\{\frac{1}{h^2}\left(\frac{\mathrm{d}r}{\mathrm{d}t}\right)^2 - \frac{1}{h^2}v^2\right\}. \tag{6.18}$$

The potentials in equation (6.17) can be compared with their Lagrangian equivalents in equations (6.10) and (6.9). Note also that the form of the law is being used in which $h = \sqrt{2}\,c$. Lévy calculated the perihelion advance to be, to first order in e,

$$\delta\varpi = \frac{f\mu}{h^2 a}(1+\alpha)nt \tag{6.19}$$

For $\alpha = 0$ one obtains the Weber value (equation (6.14)) and for $a = 1$ one obtains the Riemann value which is also the same as the Gauss value found by Tisserand. Lévy used Tisserand's value of the Weberian advance (with $h = 3000\,000$ k s^{-1}) to show that

$$\delta\varpi = (1+\alpha) \times 14.4''. \tag{6.20}$$

Lévy put $\alpha = 1.63$ or $\alpha = 5/3$ to obtain the desired $\delta\varpi = 38''$. He should by this time have used Newcomb's 1882 value of 43″ which is represented by $\alpha = 2$. Lévy's procedure is not very satisfactory, not just for its *ad hoc* nature but because he was mixing two different laws, which though they were similar had different *a priori* backings.

At the end of this paper Lévy contrasted his own formula with that of Clausius, but the distinction that he drew was that the Lévy–Weber–Riemann laws presupposed the hypothesis of two fluids. Weber had assumed this hypothesis (Fechner's hypothesis) in his 1846 derivation of the force law from Ampère's law, but in his *Treatise* Maxwell had shown that it was sufficient but not necessary for the derivation to hold. However, in 1877 Clausius showed that if Fechner's hypothesis was replaced by a one-current hypothesis in which one type of charge moved while the other remained at rest, a force would be exerted by a current-carrying wire on a charge at rest according to Weber's law. However, the force was of the order $(w/c)^2$ where w is the velocity of the current, and in 1880 von Ettingshausen showed by using the recently discovered Hall effect that a typical current velocity was of the order of millimetres per second so that the Weber force was negligible. Weber's law could then be adapted to a one-fluid hypothesis. The Clausius law did not suffer from Helmholtz's 'negative particle' objection, a plus factor for it, but it did require absolute velocity, a notion which was under attack at this time (see Mach 1960, pp. 290–7). The first person to apply it as a gravitational law was Oppenheim (1895) from whom Tisserand obtained the results which he mentioned in his *Mécanique céleste* of 1896, though they were not given explicitly. Oppenheim noted that Clausius's law did not give equality of action and reaction, and so assumed that the motion of the centre of mass of the planet–Sun system was moving in a uniform and rectilinear manner. The components of this motion then entered into the secular perturbations of the orbital elements, as well as forming a factor multiplying the gravitational constant. Of the secular perturbations, that of the perihelion dominated as it was inversely proportional to the eccentricity and was given as

$$\frac{\mathrm{d}\varpi}{\mathrm{d}t} = \mathrm{B}_1 \frac{m_0 - \mathrm{m}_1}{m_0 + m_1} \frac{n^2 a}{c^2 e} \tag{6.20a}$$

where B_1 is a linear function of the motion of the planet–Sun system, m_0 and m_1 are the masses of the Sun and the planet respectively, n, a, and e are the mean motion, semi-major axis, and eccentricity of the planet respectively, and c is the velocity of propagation of gravitational action. The quantity B_1 can therefore be fitted *ad hoc* to give the required perihelion motion for Mercury. Comparing equation (6.20a) with Weber's formula (equation (6.14)), one finds that since in equation (6.14)

$$(\delta\varpi)_{\text{Weber}} = \frac{f\mu}{h^2}\frac{n}{a}\delta t = \frac{n^3 a^2}{h^2}\delta t$$

one has, putting $c = h$,

$$(\delta\varpi)_{\text{Clausius}} \approx B_1 \frac{n^2 a}{c^2 e}\delta t$$

$$= \frac{B_1}{nae}(\delta\varpi)_{\text{Weber}}. \tag{6.20b}$$

For Mercury $(\delta\varpi_{\text{Weber}}) \approx 7''$, and we need

$$(\delta\varpi)_{\text{Clausius}} \approx 42'' \text{ (per century).}$$

Thus we want $B_1/nae=6$ or $B_1=6nae \approx 0.03$ AU per day. This corresponds well to the actual motion of the Sun (0.01 AU per day) but this value must also be close to that for the centre of mass of the Sun and Venus. Substituting back into equation (6.20b) for Venus gives

$$(\delta\varpi)_{\text{Clausius}} = \left(\frac{0{\cdot}03}{1{\cdot}6\times 2\pi/360\times 0{\cdot}7\times 0{\cdot}006}\right)\times 1''.32$$
$$\approx 390'' \text{ per century.}$$

This is far too large, so B_1 must be taken small enough to give no perihelion motions. Oppenheim did not give this calculation, but he did conclude that Clausius's law could satisfy the planetary motions as well as the inverse square law, i.e. that it could not be used to give the required advance in Mercury's perihelion. Additionally the constant of gravitation was dependent on the motion of the centre of mass of the attracting bodies, though this dependence was small being

$$G_{\text{Clausius}} = G\left(1+\frac{v^2}{c^2}\right)$$

for a velocity v of the centre of mass and c typically the same as the velocity of light. Oppenheim's investigation is the only one that we have found that discusses the gravitational use of Clausius's law. He also noted the work of Lévy and commented without elaboration on the simple way in which the perihelion motion could be obtained by his method.

In a second paper Lévy (1890*b*) summarized the alternatives for α. With $\alpha=0$ one obtained Weber's formula and the general equations of Kirchoff, with $\alpha=\frac{1}{2}$ one obtained Maxwell's equations, and with $\alpha=1$ Riemann's formula and the theory of F. Neumann were obtained. Lévy then pointed out what has been argued here, that Weber's $h=\sqrt{2}\,c$ should have been used and not $h=c$. This had the effect of halving the calculated perihelion motions, so that as a fraction of the required anomalous motion the law gave the following results: Weber ($\frac{3}{16}$), Gauss or Riemann ($\frac{3}{8}$), and Maxwell ($\frac{1}{4}$), with $\alpha=4$ required to produce the full value. Lévy concluded

> Nothing in the known facts is opposed to one adopting this figure [of α]; but nothing either is of the type to make it particularly likely. (Lévy 1890*n*, p. 742)

However true this was, Lévy's procedure was just unacceptable *a priori.*

6.4. The Maxwellian approach

The Maxwell approach was rare in gravitation. As far as electrodynamics was concerned this theory was dominant, but not sufficiently so to stop any confirmed action-at-a-distance theorist from developing that line of thought.

Indeed the line had yet to reach its peak with the Liénard–Schwarzschild formula and the theory of Ritz. Yet there was a little interest in gravitation along Maxwellian lines.

Maxwell (1865) himself had considered the possibility of tracing gravitation to the action of a medium. However, since the gravitational force between like bodies is attractive rather than repulsive, as in the electrostatic or magnetic cases, and there are no negative masses, the energy of the gravitational field decreases when two bodies approach each other. Since we cannot allow negative energy, the energy of the field when no gravitating force is present must be extremely high, in fact sufficiently high for the energy still to be positive when the greatest possible gravitational force in the universe is present. Ascribing a vast intrinsic energy to the ether was a step Maxwell could not take:

> As I am unable to understand in what way a medium can possess such properties, I cannot go any further in this direction in searching for the cause of gravitation. (Maxwell 1865, p. 493)

To an ether theorist this might sound an empty objection since the density ascribed to the ether could be surprisingly high. Sir Oliver Lodge in 1907 estimated a value 10^{10} times that of platinum, with an intrinsic energy of 10^{33} ergs cm^{-3}. The latter value, as de Tunzelmann (1910, p. 425) pointed out, allows a gravitational force 10^{10} times that at the surface of the Sun. Difficulty might be encountered with the Seeliger argument that an infinite universe implied infinite gravitational forces, though this can be evaded on several counts as we have seen in Chapter 4.

Lorentz (1900) gave a Maxwellian interpretation to the Zöllner–Mossotti theory. Whereas they had taken the attractive force between like charges to be greater than the repulsive force between unlike charges, Lorentz took the disturbances produced in the ether by positive ions to be slightly different from those produced by negative ions. Forces between like ions were then slightly different in magnitude from those between unlike ions, as in the older theory, but the disturbances were represented by Maxwell's equations with a velocity of propagation equal to that of light. The force formula, like that of Clausius, contained absolute velocities, was non-central, and gave secular variations to all orbital elements except the mean distance. The secular advance in perihelion for Mercury was evaluated at 1″.4 per century:

> Hence we conclude that our modifications of Newton's law cannot account for the observed inequality in the longitude of the perihelion—as Weber's law can to some extent do—but that, if we do not pretend to explain this inequality by an alteration in the law of attraction, there is nothing against the proposed formulation. (Lorentz 1900, p. 570)

This Lorentz theory was taken up by Gans (1905) by Wacker (1909), a student of Gans, and again by Gans (1912). It would seem on this type of theory that gravitational action would be affected by the material through which the action passed. In the extreme case a gravitational screen might have been possible as it was in electrostatics. Gans claimed to have removed this possibility by assuming the positive ions in a conductor to remain fixed and only the negative ions to move and by identifying the gravitating mass with the positive ions. One troublesome effect on this theory was that, whereas an accelerating charge radiated energy, an accelerating mass sucked in energy from the ether though the quantity was less by a factor equal to the gravitational constant. Gans seemed not to be too perturbed:

> It is noteworthy how much the views on the ether have changed. Whilst earlier people were troubled whether the ether would not dampen the planetary motions by friction, we now put the question the other way round, whether it will not increase the existing accelerations through the expenditure of energy. (Gans 1912, pp. 83–4)

The tacit inference was that this objection was not to be taken as fatal, just as the earlier objection had not been fatal. Having developed the force formula Gans determined the perihelion advance to be $\Delta\varpi = \pi v^2/c^2$ where v is the velocity of the planet relative to the Sun. For Mercury this was equivalent to 6″.9 per century, much the same as for Weber's law. Having initially posed the question 'Is gravitation of electromagnetic origin?', Gans ended with the answer '*non liquet*'—nothing is proven. A Maxwell approach (without the Zöllner base) has been investigated more recently (Coster and Shepanski 1969). They found a perihelion advance for Mercury of 7″ per century.

6.5 The theory of Ritz

The most advanced action-at-a-distance theory was published just after the turn of the century by the young Swiss physicist Walther Ritz (1908). He died in 1909 at the age of 31, and suffered continual ill health throughout his short life. He had been a fellow student of Einstein in Zurich and did much of his work in Göttingen. Ritz's gravitational law differed from the ones we have been considering in that it could be fitted *ad hoc* to give the full perihelion advance of Mercury and also in that Ritz hoped to produce a grand theory covering gravitational and electromagnetic phenomena. In 1909 he gave the outline of a theory in which gravitational action resulted from the residual effect of high-order terms in the force law of the electrodynamic theory itself. Ritz died just after this paper appeared and the later idea has not been developed.

The theory of Ritz is a ballistic one, i.e. action is propagated by fictitious particles with a velocity which is the sum of the velocity of light and the velocity of the source. The theory (Ritz 1908) resulted in the fundamental force given in equation (6.13). Ritz wrote this in the gravitational form:

$$F_x = \frac{\mu m}{r^3} x\left[1+\frac{3-k}{4c^2}\left\{\left(\frac{\mathrm{d}x}{\mathrm{d}t}\right)^2+\left(\frac{\mathrm{d}y}{\mathrm{d}t}\right)^2\right\}-\frac{3(1-k}{4c^2}\left(\frac{\mathrm{d}r}{\mathrm{d}t}\right)^2\right]$$
$$-\frac{\mu m(k+1)}{2c^2r^2}\frac{\mathrm{d}x}{\mathrm{d}t}\frac{\mathrm{d}r}{\mathrm{d}t}. \tag{6.21}$$

The constant k in equation (6.21) is undetermined. Ritz obtained the following values for the centennial perihelion advances: Mercury, $(k+5)\times 3''.6$; Venus, $(k+5)\times 0''.7$; Earth, $(k+5)\times 0.3''$. Fitting the value to Mercury's 41″ gave $k=6.4$, which yielded 41″ for Mercury, 8″ for Venus, and 3″.4 for Earth. The centennial advance for the lunar perigee may be shown to be 0″.09, which was consistent with Brown's lunar theory. Ritz thought the advances for Venus and Earth were too great, though they were much the same as obtained in General Relativity. Recently Fox (1965) suggested a procedure for evaluating k that is not *ad hoc*; if the theory were extended to include a treatment of the fine structure of the fine spectrum of hydrogen, k could then be evaluated. However, this does not immediately help for there is no reason why the constant in the electrodynamic law should have the same value as that in the gravitational law. The expression for the deflection of light in a gravitational field is independent of k so this could not have been used.

Ritz (1909) put forward some suggestions which brought gravitation into the sphere of electrodynamic action, but even this did not help. The gravitational terms were high-order terms whose coefficients remained unknown, and the theory died with Ritz in this same year. Ritz considered the atom to be made up of positive charges revolving round a central negative charge. The electrostatic charge between two neutral bodies was zero, but the velocity and acceleration terms were not. He used relative velocities and classical kinematics, and criticized Lorentz's two theories—the 1900 Zöllner-type theory and a Lorentz-invariant theory (see Chapter 7)—for giving zero forces on account of their use of absolute velocities and non-classical kinematics respectively. As Gans had found, Ritz encountered the possibility of gravity screens and also required the positive charges to be fixed.

The second-order terms in the force law depend on the square of the velocities of the charges. This gave the possibility of temperature-dependent terms which according to Ritz were 'contrary to observation'. In fact few experiments had been performed which tested the commonly held assumption that gravitational action was independent of temperature. In 1916 such a temperature dependence was claimed to have been found by P.E. Shaw at Nottingham (Shaw 1916*a*). He did not refer to Ritz but did refer to Mie's theory of matter (Mie 1913), to a wave theory of Morozov (1908), and to some ideas of Bohr (1913) for some theoretical basis for this dependence. Previously, in 1905, Poynting and Phillips had found slight evidence for a change in weight of a body when its temperature was changed, and in 1906

Southern obtained a similar result. Shaw's innovation was to change the temperature of the *larger* mass in a modified form of Boys' apparatus over a range of 20–240°C, and he obtained on rather slender evidence a tentative modification of the inverse square law of the form

$$F = G(1+\alpha\theta)\frac{\mathrm{Mm}}{\mathrm{d}^2} \tag{6.22}$$

The temperature θ in equation (6.22) refers to the larger body of mass M; the temperature coefficient α was found to be $+1.2\times 10^{-5}$ per °C.

There followed a long discussion on Shaw's results in the correspondence pages of *Nature* (Larmor 1916, 1917; Shaw 1916*b*, 1917; Barton 1916, 1917; Lindemann and Burton 1917; Todd 1917; Lodge 1917; Thomas 1917). Larmor thought that disagreement with experience would be obtained in the case of a comet which was heated on approaching the Sun. If its inertial mass increased, its velocity would have to decrease for conservation of momentum to hold and its orbital motion would be disturbed. Shaw replied that equation (6.22) was only valid when it was the larger mass that was being heated. Another more telling point of Larmor was that according to electrodynamic theory

$$\delta m = \frac{\delta E}{c^2} = \frac{k}{c^2}\delta T \approx 10^{-35}\,\delta T$$

and that this was 'minute beyond detection' and much smaller than the values of α in equation (6.22). Shaw gave a wave theory interpretation of the temperature dependence: the temperature-dependent

> . . . attraction is due to vibration of the Faraday tubes, which are carried to-and-fro by the molecules in their vibratory motion. This is like Challis's wave theory of gravitation, whereby bodies in a vibratory medium attract one another if their phases are in close agreement. (Shaw 1916*b*, p. 401)

He obtained a force law which was the same as had been suggested by Poynting and Phillips, and which reduced to equation (6.22) for large mass M and small mass m. Lindemann and Burton wrote a joint letter criticizing various consequences of Shaw's result but praising his experimental work. The discussions ended in 1917, but the question was resolved in 1922 when the experiments were repeated by Shaw and Davy (1923) and the effect was found for all intents and purposes to have disappeared. Defects in the supporting arrangements were blamed for the 1916 result.

Rather than allow a small temperature dependence in his gravitational force, Ritz decided to choose the arbitrary constants so as to get rid of the responsible terms of the second order. Terms of higher order did not suffer from this objection:

> . . . on condition that the intra-atomic velocities are large compared to the velocities produced by thermal agitation, which is probable *a priori*. (Ritz 1911, p. 490)

Thus Ritz obtained the basic law of Newton in the fourth-order term having made the plausible assumption that the mass of a body was proportional to the number of rotating charges contained in it. The sixth-order terms would account for the perihelion advance of Mercury (and to a certain extent starlight deflection in the Sun's gravitational field) if the constants were suitably evaluated, though this would be restricted by the need to get rid of the second-order terms. Ritz also invoked the third, fifth, etc. orders as giving possible contributions to gravitation since these also had undetermined constants. He concluded:

> . . . that an explanation of the anomaly of Mercury and a determination of the gravitational constant by electromagnetic measurements could without doubt be deduced from the laws of electrodynamics, when they are known with more exactness.
>
> On either of the hypotheses, gravitation will essentially be the result of the dynamic constitution of atoms. (Ritz 1911, p. 490)

Ritz died in 1909 and his theory remained a bare skeleton. Whether one adopted this theory of just used the force law, its acceptability depended on the acceptability of the original electrodynamic theory. This electrodynamic theory was rejected on experimental grounds, but it has been argued by Fox (1962, 1965) that conclusive evidence against Ritz was obtained only in the 1960s. Fox actually had to modify Ritz's theory for originally the velocities of the fictitious particles were unchanged when they passed through a material medium. Since this disagreed with Fizeau's experiment, Fox introduced a natural modification:

> When a light wave sets into motion the charges of a medium, these in turn emit new waves whose centres move in vacuum with the velocity of the charges of the medium. (Fox 1965, p. 4)

Fox was then able to explain all optical experiments which seemed to have refuted Ritz by invoking the

> . . . so-called extinction theorem of Ewald and Oseen which shows how an external electromagnetic disturbance travelling with the velocity of light in vacuum is exactly cancelled out and replaced in the substance by the secondary disturbance travelling with an appropriate smaller velocity. (Born and Wolf 1959, p. 70)

This theorem shows that experiments concerned with the velocity of light from a particular source invariably measure the velocity of light characteristic of some intervening material. The effect of the extinction theorem can only be escaped from when 'the electromagnetic radiation [is] of a frequency so high that atomic electrons cannot follow it'. This condition is fulfilled by γ rays, and Fox took as the best evidence against Ritz's theory and for the postulate of the constancy of light with respect to a moving source experiments on 6 Gev γ rays carried out in 1964.

Ritz's theory was seen as a serious competitor to special relativity. One of the common arguments against it came from astronomy, which was de Sitter's argument from double stars, put forward in 1913 (de Sitter 1913*c*) although the basic idea had been published by Comstock (1910).

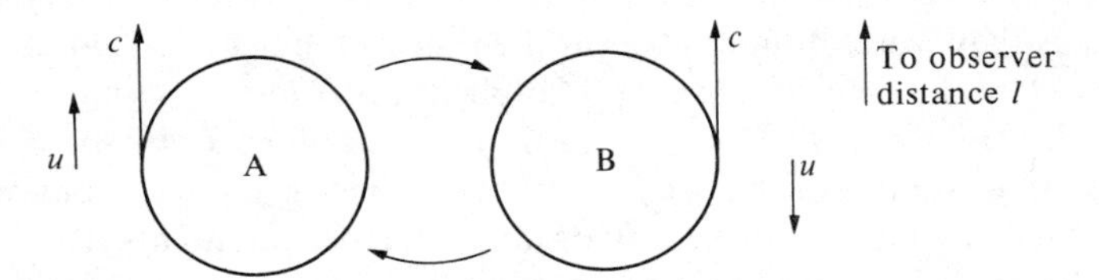

FIG. 6.1. A binary star rotating about its centre of mass.

If an emission theory of light be true, the velocity of light from the star in position A will be $c+u$, while in the position B the velocity will be $c-u$. Hence the star will be observed to arrive in position A, $l/(c+u)$ seconds after the event has actually occurred, and in position B, $l/(c-u)$ seconds after the event has occurred. This will make the period of half rotation from A to B appear to be $\Delta t - l/(c+u) + l/(c-u) = t + 2ul/v^2$, where Δt is the actual time of a half rotation in the orbit, which for simplicity may be taken as circular. On the other hand, the period of the next half rotation from B back to A would appear to be $\Delta t - 2ul/c^2$. Now in the case of most spectroscopic binaries the quantity $2ul/c^2$ is not only of the same order of magnitude as Δt but often times probably even larger. Hence, if an emission theory of light were true, we could hardly expect without correcting for the variable velocity of light to find that these orbits obey Kepler's laws, as is actually the case. This is certainly very strong evidence against any form of emission theory. (Tolman 1917, pp. 24–5)

De Sitter's argument was attacked by Freundlich (1913) and by Guthnick (1913) in the same year. Freundlich pointed out that to first order the effect of the variable velocity of light would be to produce a Keplerian ellipse whose line of apsides was in the line joining the star to the Earth with the periastron pointing away from Earth. He then added that this effect had indeed been observed, by J.M. Barr in 1908, though he admitted that this might have had some other origin, such as a cosmogonical cause connected to the Sun's position close to the centre of the Milky Way.

De Sitter (1913*d*) replied in a short note, repeating his claim and showing how, for a typical star, the influence of any variation in the velocity of light must be very small. However, Fox has used the extinction theorem to show how the force of this argument may be deflected. All extraterrestial light passes through other matter such as interstellar gas, the Earth's atmosphere, and glass in the laboratory. Moreover, it has been shown by Otto Struve that most close binary stars are surrounded by a gaseous envelope which, as it does not rotate with the two stars, would also contribute to masking the emission velocity. Struve and Huang (1957) have put forward a hypothesis specifically to account for the Barr effect. They suggested that the non-uniform distribution of periastron values was due to distortion of the radial velocity curve caused by ejected streams of gas revolving around the component stars. Nevertheless the hypothesis is not so well established as to be accepted unequivocally, and independently of this one might still use the extinction theorem together with interstellar gas etc. to save Ritz's theory.

It seems that the grounds for the rejection of Ritz's theory were not so well founded as they were thought to be at the time of the development of general relativity. Ritz himself criticized the application of electrodynamic laws as that of Weber to gravitation because they neither gave all of the perihelion anomaly of Mercury nor did they obtain (nor give the hope of obtaining) the gravitational constant from electric or magnetic measurements. Ritz's theory offered the possibility of both, and he estimated its success as 'very probable' (Ritz 1911, p. 487). There were two other empirical effects apart from the Mercury perihelion advance that eventually supported general relativity as the three 'classical' tests, these being the red shift of spectral lines emitted in a gravitational field and the deflection of light rays passing near a massive body.

The gravitational red shift according to general relatitivity is a consequence of the dilatation of time in a gravitational field. However, it may be obtained on a simple photon model. If a photon of mass m is taken from a region of potential P to a region where the potential is negligible, then the photon loses energy mP. Since according to special relativity $E=mc^2$, the photon loses energy equal to PE/c^2 so that the fractional loss $dE/E=P/c^2$, or the fractional red-shift, is $d\upsilon/\upsilon=-P/c^2$ which is the general relativistic amount. Thus all theories gave a red shift and it did not distinguish between rival theories.

The deflection of light according to general relativity is a consequence of the dependence of the velocity of light on gravitational potential. However, treating the photon as a particle also gives such a deflection. As early as 1801 J. Soldner had given a Newtonian treatment to the problem, and this results in half the general relativistic prediction of $1''.75$. Now O'Rahilly (1965, pp. 544–5) has given a general expression for an electrodynamic force law together with expressions for the perihelion advance and the deflection of starlight according to this force. Comparing the general force with Ritz's force gives the values for the coefficients in the general expression. Thus one may find that the Ritz deflection is three-quarters of that predicted by general relativity, or $1''.31$. Now although the confirmation of the general relativistic prediction was not exactly conclusive, a glance at a list of results obtained shows that in no way was a value of $1''.31$ confirmed in preference to $1''.75$ (von Kluber 1960)†. The Greenwich expedition to Principe in 1919 gave $1''.61\pm0.40$ and an Australian expedition in 1922 (Dodwell and Davidson) gave $1''.77\pm0.40$, but whereas both allow the Ritz value preference would be given to general relativity. Other values, such as the Greenwich expedition to Sobral of 1919, supported general relativity, and later expeditions where deviations were found gave results greater than $1''.75$. Thus the Ritz theory (i.e. that in which the force formula is simply used instead of Newton's inverse square law), even

† The results in the text are tabulated in detail with references on p. 58 of this reference.

if it was perhaps unfairly neglected, was effectively refuted by 1922. Nevertheless the neglect of this theory has meant that this refutation has not been noted.

6.6. The theory of Gerber

There was in the first two decades of this century a rival to both Ritz's theory and general relativity. This was a hypothesis that did not claim to give an all-embracing account of gravitation and electrodynamics but gave a simple gravitational force formula which, without the introduction of arbitrary constants or terms to be put to zero, gave a complete account of the motion of Mercury's perihelion. This was the force law of Paul Gerber.

Gerber was a Realgymnasium Professor in Stargard, Pomerania. He had been born in Berlin in 1854 and died sometime between 1902 and 1917.† He published few scientific papers but wrote for a number of popular science journals. The force law appeared in 1898 (Gerber 1898) and was given in a more detailed paper in 1902 (Gerber 1902). This paper was reprinted in 1917 at the instigation of Ernst Gehrcke (Gerber 1917). The importance of Gerber's work was not only that it gave the correct perihelion motion for Mercury but that it was also used as a weapon against general relativity in the German campaign waged against Einstein, in which Gehrcke played a leading part. Gerber saw himself in the tradition of the workers we have been discussing, and his 1902 paper reviews much of the work given above. The derivations of his force law were not at all clear, but the following gives the required result. The 'static' case has the usual Newtonian potential V_0 where

$$V_0 = m_1 m_2 / r \tag{6.23}$$

for two masses m_1 and m_2 with the gravitational constant normalized to unity. When the two approached each other at a relative velocity $\mathrm{d}r/\mathrm{d}t$, and if v is the velocity of propagation with respect to the source (and the total velocity of the potential is the sum of v and the velocity of the source) and Δt is the time taken for the potential to travel from m_1 to m_2, then $r = v\ \Delta t$ and the distance travelled by the masses is

$$\begin{aligned} r - \Delta r = r - \Delta t \frac{\Delta r}{\Delta t} &= r\left(1 - \frac{1}{v}\frac{\Delta r}{\Delta t}\right) \\ &= r\left(1 - \frac{1}{v}\frac{\mathrm{d}r}{\mathrm{d}t}\right). \end{aligned} \tag{6.23}$$

This expression should replace r in equation (6.23) but there is a further consideration. The faster the receiving mass travels, the less potential it

† When his 1902 paper was reprinted in 1917 his name was followed by a cross. Poggendorff does not give note of his death.

receives in unit time. The velocity of this mass with respect to the potential is $v - \mathrm{d}r/\mathrm{d}t$ so the potential V must be multiplied by a factor $v/(v-\mathrm{d}r/\mathrm{d}t)$, i.e. unity when $\mathrm{d}r/\mathrm{d}t = 0$. Thus we have for the general expression for the potential

$$\begin{aligned} V &= m_1 m_2 \bigg/ r\left(1-\frac{1}{v}\frac{\mathrm{d}r}{\mathrm{d}t}\right)\left(1-\frac{1}{v}\frac{\mathrm{d}r}{\mathrm{d}t}\right) \\ &= m_1 m_2 \bigg/ r\left(1-\frac{1}{v}\frac{\mathrm{d}r}{\mathrm{d}t}\right)^2 \\ &= \frac{m_1 m_2}{r}\left\{1+\frac{2}{v}\frac{\mathrm{d}r}{\mathrm{d}t}+\frac{3}{v^2}\left(\frac{\mathrm{d}r}{\mathrm{d}t}\right)^2\right\} \end{aligned} \tag{6.24}$$

to second order in v. This potential can be substituted into the standard Lagrangian equation

$$F = \frac{\mathrm{d}v}{\mathrm{d}r} - \frac{\mathrm{d}}{\mathrm{d}t}\left(\frac{\mathrm{d}V}{\mathrm{d}\dot{r}}\right)$$

to give

$$F = -\frac{m_1 m_2}{r^2}\left\{1-\frac{3}{v^2}\left(\frac{\mathrm{d}r}{\mathrm{d}t}\right)^2+\frac{6r}{v^2}\frac{\mathrm{d}^2 r}{\mathrm{d}t^2}\right\}. \tag{6.26}$$

This force law has the remarkable property that if one puts v equal to the velocity of light the perihelion advance for Mercury turns out to be 41″. For Venus the value is 8″, the same as for Ritz's law and for general relativity, and the values for the other planets and for the lunar perigee are all small enough to agree with observation. This formula seemed to have solved the problem that had blocked the acceptance of the Weber-type laws—it gave the whole of the Mercury anomaly while using the 'natural' velocity of propagation, that of light.

Gerber's law provoked little reaction on its first publication, though it appeared in Mach's *Die Mechanik* (Mach 1904, p. 201; see Mach 1960, p. 235), in Zenneck's review of gravitation (Zenneck 1903), and in de Tunzelmann's text book (de Tunzelmann 1910); it was not mentioned in Poincaré's 1906/7 lectures on the limits of Newton's law (Poincaré 1953). It was brought to life again by Ernst Gehrcke when Einstein's General Theory of Relativity appeared. The expression for the advance in the perihelion of a planet obtained from Gerber's law is

$$\delta\varpi = \frac{24\pi^3 a^2}{T^2 c^2 (1-e^2)} \tag{6.27}$$

and this is exactly the same as that obtained in general relativity. Gehrcke (1916) pointed this out in a paper criticizing the theory of Einstein, and claimed priority for Gerber. Gehrcke's role is important, for he was one of the leading members of the *Arbeitsgemeinschaft Deutcher Naturforscher* set up to try and discredit Einstein and the relativity theory. Einstein dubbed them the 'Anti-relativity Company', and they worked at a popular level. Their leader was Paul Weyland, of whom Ronald Clark write that he was 'a man entirely unknown in scientific circles and of whom, over the years, nothing was discovered' (Clarke 1971, pp. 256–9). A certain scientific respectability was lent by the membership of Philip Lenard, a Nobel Prize winner who gained that distinction for experimental work and who later wrote the antisemitic *Deutsche Physik* (Lenard 1936–7). However, the presence of Gehrcke shows that there was at least some scientific basis to this seemingly irrational enterprise, for it did appear that the empirical support for general relativity lay solely in the perihelion advance of Mercury, this being before any eclipse expeditions had been completed and at a time when solar spectral red shifts were equivocal. Here was a theory which gave the same result for this all-important anomaly and yet seemed so much simpler. Owing to Gehrcke's efforts the 1902 paper of Gerber was republished in 1917 in *Annalen der Physik*, and this time a response was obtained. The first to attack was Seeliger (1917*a*), the man responsible for the zodiacal-light hypothesis for explaining Mercury's perihelion that was the subject of Chapter 4. Seeliger claimed that Gerber's calculation was based on an elementary mistake, though it is evident that it was Seeliger who was mistaken. He also objected to velocity-dependent potentials, but that is not the point here. Seeliger noted that the Weberian potential

$$V = \frac{\mu}{r}\left\{1 - \frac{1}{c^2}\left(\frac{\mathrm{d}r}{\mathrm{d}t}\right)^2\right\} \tag{6.28}$$

gave the force

$$R = \frac{\mu}{r^2}\left\{1 - \frac{1}{c^2}\left(\frac{\mathrm{d}r}{\mathrm{d}t}\right)^2 + \frac{2r}{c^2}\frac{\mathrm{d}^2 r}{\mathrm{d}t^2}\right\}. \tag{6.29}$$

Gerber had obtained his force formula (equation (6.26)) from his potential (equation (6.24)) using the Lagrangian equation (equation (6.25)). Seeliger now expected equation (6.25) to be used for the Weberian potential (equation (6.28)), which results in

$$R = \frac{\mu}{r^2}\left\{1 + \frac{1}{c^2}\left(\frac{\mathrm{d}r}{\mathrm{d}t}\right)^2 - \frac{2r}{c^2}\frac{\mathrm{d}^2 r}{\mathrm{d}t^2}\right\}. \tag{6.30}$$

The force in equation (6.30) gives a retardation of the perihelion of the same amount as equation (6.29). However, if the Lagrangian equations are to be used one must see the equivalent Weberian potential (equation (6.10)), and the combination of equations (6.10) and (6.25) gives equation (6.29) as required. To obtain equation (6.29) from equation (6.28) one just differentiates with respect to r, equation (6.28) being the same as the form of Weber's potential, but this I think accidental and if a generalized potential is to be used it should be in the Lagrangian form. If the potential has no velocity-dependent terms, equation (6.25) reduces to the normal $F = \mathrm{d}V/\mathrm{d}r$.

Seeliger repeated his claim later in 1917 (Seeliger 1917b) when he allowed that one could use what Carl Neumann had called an effective potential but that this was not what Gerber had meant. Now it is possible that Gerber himself did not know what he meant, except that this was a velocity-dependent potential with a velocity of propagation equal to that of light. Seeliger was impressed by the remark he quoted at the end of his paper from Neumann (1896, p. 245) in which he had written that transmission of the effective potential was

> . . . a completely transcendental concept, essentially different from the propagation of light or heat. (see O'Rahilly 1965, p. 186)

However, in an earlier passage one can see that according to Neumann (1896) the transcendental nature is not in the propagation itself but in the nature of the interaction.

> We treat this potential as a stimulus to motion or, to use a better expression, as a command which is given and emitted by one point and is received and obeyed by the other; we assume that this command requires a certain time in order to travel from the place of emission to the place of reception. (see O'Rahilly 1965, p. 186)

The reception of the potential is an important part of Gerber's theory, for his formula (equation (6.24)) was obtained by considering the effect the finite time of propagation had, first on the distance to be used and second on the amount of potential received. The latter can be interpreted in terms of Ritzian fictitious particles, i.e. the faster the passive body moves towards the active one the more of these particles it 'collects', or the faster the active body moves the faster the particles travel and the more there are available per unit time. However, the details of this reception are difficult to imagine.

In 1918 Gehrcke published a paper supporting the ether views and criticizing Seeliger (Gehrcke 1918) which drew replies from Einstein (1918) (on the ether) and Seeliger (1918) to which Gehrcke (1919) responded in turn, suggesting that Seeliger's opposition to Gerber was not neutral as Seeliger had his own explanation of the Mercury advance.

There had been other responses to Gerber's republished paper. Oppenheim, who had published a review of gravitation theories using finite velocities of propagation in 1895, republished an extract from a 1903 paper on the Newtonian gravitational law (Oppenheim 1917). He showed how, given a potential of the form

$$P = \frac{x^2 m_1 m_2}{r\{1-(1/c^2)(\mathrm{d}r/\mathrm{d}t)^2\}^{\lambda}} \tag{6.31}$$

one obtains a perihelion advance of

$$\Delta\varpi = \frac{\lambda(\lambda+1)}{2}\frac{n^3a^2}{c^2}. \tag{6.32}$$

One obtains Weber's law with $\lambda = 1$, and if c is the velocity of light this yields $\Delta\varpi = 13''.65$. (This is incorrect, for as has been shown Weber's law gives half of this; compare Weber's potential (equation (6.4)) with equation (6.31) for $\lambda = 1$.) One obtains the full perihelion motion of $41''.25$ for $\lambda = 2$ or $\lambda = -3$. He showed that Lévy's potential combining the Weber and Riemann potentials was equivalent to the case for $\lambda = 2$, and that since Gerber's potential was also the $\lambda = 2$ case all Gerber had to do was to provide 'a physically plausible basis' for this form:

> To what extent the proof, as he carries it out, is sound and satisfies the physicists, I abstain from each decision on that.

A third response was given by von Laue (1917), in which he displayed a reluctance to accept that action at a distance could be coupled with propagated action.

Not everyone abstained with Oppenheim. von Gleich (1923) noted the identity of Gerber's perihelion advance with that of Einstein but added

> But the proof of his potential can not be considered as sound. (von Gleich 1923, p. 230)

von Laue had also written,

> And what otherwise Gerber brings forward as physical considerations appears unintelligible. . . . We cannot therefore recognise Gerber's work as physical explanation. (von Laue 1920, p. 736)

von Laue also inferred that Gerber had merely worked backwards on the lines that Oppenheim had suggested. Bucherer also referred to Gerber:

> It is the merit of Gerber to have tried the calculation of the orbit of Mercury based on the assumption that gravitational action propagated with the velocity of light. He reached—of course by means of false inferences—a formula for the perihelion motion identical with that of Einstein. (Bucherer 1922, p. 5)

However, if the vague nature of Gerber's work forced vague responses, effort could have been saved by pointing out that Gerber's law was in fact refuted, and this on two counts.

It has been shown how this law leads to an identical expression for the perihelion motion as does general relativity. However, Gerber's theory is a gravitational theory and says nothing outside that sphere. Hence for a complete physical picture it must be joined by, among other things, a theory of electrodynamics. At the time of the advent of the general relativistic theories this would have been either Lorentz's theory or special relativity, but as we shall see in the next chapter both gave the same relationship for the variation of mass with velocity. This dependence, when applied within 'Newtonian' theory, resulted in a 7″ advance in the perihelion of Mercury.† We have already seen in Chapter 4 how in 1913 de Sitter showed that this 7″ could easily be accommodated within the then standard Newtonian explanation of Mercury's anomaly, Seeliger's zodiacal light hypothesis, by simply decreasing slightly the density of the matter ring. This relativistic advance must be taken into account when applying gravitational theory. This is done tacitly in general relativity because this theory already contains special relativity. It would be done tacitly in the uncompleted 1909 Ritz theory because that was a complete theory covering electrodynamics as well as gravitation, and the mass variation etc. is found in the electrodynamic theory of Ritz (O'Rahilly 1965, p. 620). In the 1908 use of the Ritz force law purely as a gravitational law the electrodynamic advance would have had to have been added, but since the fitting of the anomaly was *ad hoc* this would not have refuted the law. However, the apparent virtue of Gerber's law in having no *ad hoc* constants now turns against it, for we must add the relativistic advance in this case also which results in an overall advance of about 49″. Whereas a small deficit in the advance could conceivably have been ascribed to small solar oblateness or intra-Mercurial matter, a small excess had no hope of being explained. No one would claim that the Sun was actually prolate, and attacks on the validity of the 42″ advance claimed if anything that it was too large. Gerber's law, therefore, did not give the correct perihelion advance.

The second count on which Gerber's law may be seen to be refuted is the deflection of light rays, though this was at times rather equivocal.

We have already discussed the deflection of light in the gravitational field of the Sun on Ritz's theory. We treated the light ray as composed of particles passing near a massive body, and we can repeat the procedure for Gerber's law. Gerber's potential is

$$V = \frac{GMm}{r(1-\dot{r}/c)^2} = \frac{\mu m}{r}\frac{1}{(1-\dot{r}/c)^2} \tag{6.33}$$

and so his Lagrangian L is given in terms of the total energy E and V as

$$L = mc^2\left(1-\frac{1}{\gamma}\right)+\frac{\mu m}{r}\left(1-\frac{\dot{r}}{c}\right)^{-2}$$

or

$$L = mc^2-\frac{mc^2}{\gamma}+\frac{\mu m}{r}\left(1+2\frac{\dot{r}}{c}+3\frac{\dot{r}^2}{c^2}\right)$$

where

$$\gamma = (1-v^2/c^2)^{-\frac{1}{2}}. \tag{6.34}$$

† Newtonian theory is of course incompatible with Special Relativity—according to it, for instance, gravitational action is propagated in infinite speed. Gerber's theory is in this respect compatible.

By means of the usual Lagrangian treatment we obtain the equation of motion

$$\ddot{u}+u-\frac{\mu}{h^2}+\frac{3\mu}{c^2}\dot{u}^2+\frac{6\mu}{c^2}u\ddot{u} = 0 \tag{6.35}$$

where $u=1/r$ and h is the constant of integration from the law of areas $(\mathrm{d}/\mathrm{d}t)(r^2\dot{\theta})=0$. The solution of this is

$$u = \frac{\mu}{h^2}\left\{1+\varepsilon\cos\left(1-\frac{3\mu^2}{h^2r^2}\right)\theta\right\} \tag{6.36}$$

which is a rotating ellipse with an advance of perihelion of $3\mu^2/h^2c^2$ per revolution, as is found in general relativity.

To calculate the deflection of light we consider a particle approaching from infinity. For this we put $h=-\infty$, so that from equation (6.35)

$$\ddot{u}+u+3\frac{\mu}{c^2}\dot{u}^2+6\frac{\mu}{c^2}u\ddot{u} = 0$$

or

$$\ddot{u}+u = -\frac{3\mu}{c^2}\dot{u}^2-\frac{6\mu}{c^2}u\ddot{u}. \tag{6.37}$$

The solution of $\ddot{u}+u=0$ is $u=(1/R)\cos\theta$. The solution of the whole equation is then

$$u = \frac{1}{R}\cos\theta+\frac{3\mu}{R^2c^2}\sin^2\theta$$

or

$$\frac{1}{r} = \frac{1}{R}\cos\theta+\frac{3\mu}{R^2c^2}\sin^2\theta$$

or

$$x = R\mp\frac{3\mu}{Rc^2}\frac{y^2}{(x^2+y^2)^{\frac{1}{2}}}$$

where $x=r\cos\theta$ and $y=r\sin\theta$. Thus R is the distance of closest approach, or for our purposes the radius of the Sun. On both sides of its passage past the Sun the ray deviates an amount $3\mu/Rc^2$ from the straight line $x=R$ (consider $y\gg x$), so that the total deviation is $6\mu/Rc^2$. This quantity is $\frac{3}{2}$ that obtained in general relativity, and is thus 2″.62.

This would now be considered a clear refutation of Gerber's law. Earlier eclipse results were not conclusively supportive of general relativity (von Kluber 1960). The 1919 Sobral expedition (1″.98 ± 0″.16) did confirm it, in the sense that the main rivals, the theory of Nordström and the 'Newtonian' theory, gave no deflection and 0″.87 respectively. This also would have favoured general relativity over the 1″.31 of Ritz and the 2″.62 of Gerber. However, later results showed values greater than the Einstein prediction, foremost of which was the 1929 result of Freundlich which gave $2''.24 \pm 0.10$. This was equivocal between Einstein and Gerber. A Russian result in 1949, of $2''.73 \pm 0.31$, gives preference to Gerber, but later results failed to confirm such a high value and general relativity was not taken to be refuted.

Between the Sobral result and Freundlich's expedition eclipse measurements favoured Einstein. Since Einstein's supporters did not concern themselves with Gerber's work it is not surprising that this result is not found in the literature—Gerber partisans would not proclaim their inferiority—but one might have expected some notice of it in 1929. Nevertheless Gerber's law faded from view rather than being publicly refuted.

6.7 Some later attempts

Other occasional references to Weber-type laws can be found at this time. At the Royal Astronomical Society meeting of December 1919 when a discussion on relativity took place, Lindemann announced that he and 'Major Griffith'† had been investigating laws of this type and were going shortly to publish a paper on a potential $m_1 m_2 f/r(1 + r\dot{\varnothing}^2/c^2)$ (Fowler *et al.* 1919, pp. 115–6), but such a paper was never published. The potential bears a resemblance to that of Riemann, in which the term in brackets is $(1 + r^2 \dot{\varnothing}^2/c^2 + \dot{r}^2/c^2)$ and Lindemann claimed to obtain

> . . . identically the same equation as Einstein for Mercury's orbit, and under certain fairly plausible assumptions for the deflection of light. . . . the fact that such a simple and almost necessary correction to Newton's law may explain the facts seems worthy of notice. (Fowler *et al.* 1919, p. 116)

Some French and Belgian papers appeared in the 1920s on the application of general electrodynamic potentials but none recommended their adoption; rather they showed the various potentials that gave the required perihelion shift (Bertrand 1922, Dehalu 1926, Swings 1927).

Surdin (1962) has obtained a potential of this type which has been claimed to fit a more general approach to physics—the ether approach of Prokhovnik (1962). Surdin's derivation of his potential is as follows:

> It is assumed here that when in motion [the mass] m emits, in all directions, a gravitational wave, propagating in free space with the velocity of light c; this wave reaches the attractive centre M, is reflected and received back by m.
>
> This note is based on the assumption that the information concerning the gravitational potential created by M is transported by this wave. Consider two positions, A and B, of the mobile point m, such that the time spent by m to get from A to B is equal to the time of propagation of the gravitational wave from A, via M, to B. Let Δt be this time of propagation, one has

† I do not know for certain who this was, but of the two Griffiths who were FRAS, one, I.O. Griffith, was mentioned in Lindemann's biography (Birkenhead 1961).

$$\Delta t = \frac{2r}{c} - \dot{r}\frac{\Delta t}{c}$$

hence

$$\Delta t = \frac{2r}{c(1+\dot{r}/c)}.$$

During the time Δt the distance between m and M has increased by a quantity

$$\dot{r}\Delta t = \overline{MB} - \overline{MA} = \frac{2r\dot{r}}{c(1+\dot{r}/c)}.$$

For mobile m the apparent gravitational potential at B is the Newtonian potential existing at A, i.e.

$$\Phi_B = \frac{GM}{r - 2r\dot{r}/c(1+\dot{r}/c)} = -\frac{GM}{r}\frac{1+\dot{r}/c}{1-\dot{r}/c} \qquad [(6.38)]$$

where G is the constant of gravitation. The final result $[\delta\varpi]$ is the same for a potential energy of the form

$$\Phi_{B1} = -\frac{GM}{r}\frac{1-\dot{r}/c}{1+\dot{r}/c}$$

and it is still the same for a potential energy

$$\Phi = \tfrac{1}{2}(\Phi_B + \Phi_{B1}) = -\frac{GM}{r}\frac{1+\dot{r}^2/c^2}{1-\dot{r}^2/c^2}.$$

Surdin's potential (equation (6.38)) can be written to second order in $1/c$:

$$\Phi = -\frac{GM}{r}\left(1 + \frac{2\dot{r}}{c} + \frac{2\dot{r}^2}{c^2}\right) \qquad (6.39)$$

which gives, using the Lagrange equation (6.25), the force

$$R = \frac{GMm}{r^2}\left(1 - 2\frac{\dot{r}^2}{c^2} + 4\frac{r\ddot{r}}{c^2}\right). \qquad (6.40)$$

The coefficient of the term $(\dot{r}/c)^2$ is the one which determines the magnitude of the perihelion motion, which is an advance if the sign is negative. Comparison with the pseudo-Weber force law (equation (6.29)) and the Gerber force law (equation (6.26)) shows that Gerber gives the full motion (the coefficient is 3), Surdin gives $\frac{2}{3}$ (the coefficient is 2), and pseudo-Weber gives $\frac{1}{3}$ (the coefficient is 1), or 40″.95, 27″.30, and 13″.65. Surdin used a Lagrangian method to show this. However, he also introduced the relativistic effect of variable mass, adding an extra 7″ to the advance, which gives about 35″ per century for Mercury or five-sixths of the amount obtained by general relativity. Though Prokhovnik felt that this was satisfactory, the only way of making up the deficit would be by solar oblateness, and recent measurements of that by R.H. Dicke are at best contentious.

Surdin also showed that his potential gave the same result as general relativity for the deflection of starlight in the gravitational field of the Sun. Though action-at-a-distance theories still occasionally appear, the theory of general relativity is so well entrenched that discussions on gravitational theories now take place using its terms. The development of general relativity and its relation to the anomalous perihelion advance of Mercury will be considered in the next chapter, and it will be noted there how little the development had to do with the topics of this chapter. The theories of this chapter culminated in the theories of Gerber and Ritz; that of Gerber, of dubious scientific pedigree, was ignored apart from a few partisans of questionable intent, and that of Ritz, a theory of high pedigree but containing a certain amount of arbitrariness and a gravitational theory of questionable worth, was hardly taken up. Neither fully completed in terms of the three classical tests of general relativity, which was unfortunate in view of their refutation by the measurements of light deflection.

7

GENERAL RELATIVITY AND THE ANOMALOUS ADVANCE IN THE PERIHELION OF MERCURY

7.1. General relativity

Einstein's general theory of relativity was published at the end of 1915. Its eventual acceptance as the gravitational theory to succeed Newtonian theory rested on three pieces of evidence that have become known as the classical tests of general relativity. The three predictions supported in these tests were:

1. a perihelion advance of planetary orbits which for Mercury agreed with Newcomb's value of about 42″ per century;
2. a bending of light rays near a massive body, measured using starlight deflected by the Sun during a total eclipse (for a historical review see von Klüber (1960));
3. a red-shift measured in light emitted by a massive body such as the Sun (for a historical review see Forbes (1961)).

As Einstein worked to produce a relativistic gravitational theory it was clear that it would give at least a certain amount of the anomalous perihelion advance of Mercury, purely on the grounds that any alteration of the Newtonian law would do so. It will be part of our purpose to see just how far it was possible to estimate the likely advance and whether the value could have been used to judge the success of the developing theory. The discussion of general relativity and its rivals will be mainly on this empirical level. We have divided Einstein's theory into four constituent parts in order to facilitate this treatment: (1) special relativity; (2) non-Euclidean geometry; (3) the principle of equivalence; and (4) the principle of covariance.

Einstein did not work in isolation, and the years up to 1915 were full of debate and argument between a number of physicists concerned with developing a new gravitational theory. Two of these of great importance were Max Abraham and Gunnar Nordström. Abraham was born in 1875, studied with Planck in Berlin and became a Privatdozent in Göttingen in 1900. He had a sharp tongue which is said to have caused him to be passed over for academic appointments. Abraham eventually obtained a chair of rational mechanics in Milan, from where he returned to Germany in 1914. He became a professor of physics at a technische Hochschule in Stuttgart, and died in 1922 just after being appointed to a chair of theoretical physics in Aachen. Nordström was born in 1881 in Helsinki. He studied in Göttingen between 1906 and 1907, became a Dozent at Helsinki University from 1910 to 1918 and then professor at the technische Hochschule from 1918 until his death in 1923. Einstein's career had a slow start. It took time for his paper introducing special relativity in 1905 to achieve a favourable reception, and he did not obtain an academic appointment until 1909 when at the age of 30 he was made an associate professor at Zürich University. He moved to Prague in 1911, returned to Zürich in 1912 as full professor and then moved to Berlin in 1914 as Director of the Institute of Physics at the Kaiser Wilhelm Institute.

Although the early reception given to Einstein's special relativity was not wholly favourable, it quickly became a standard reference point among the physicists developing gravitational theories. Newtonian theory was incompatible with special relativity because its instantaneous propagation of action violated the relativistic principle that the velocity of light was the maximum velocity of propagation. A number of scientists worked on giving Lorentz-invariant adaptions of the Newtonian law which will be discussed in §7.5. The more progressive theories started from basic considerations. Abraham (1914) gave a simple illustration of a distinction between two groups of gravitational theories. The total energy E of a body of mass m is a function of the gravitational potential Φ at the point where it is situated. The energy E appears in the well-known relativistic law $E=mc^2$ where c is the velocity of light. If E is a function of Φ then one of m and c must be a function of Φ. In his early gravitational theories Einstein put $c=f(\Phi)$, so that special relativity was only valid in small regions of space-time in which the gravitational potential was constant (to give a constant value for c) or far from matter where the potential would be zero. Abraham also chose $c=f(\Phi)$ but he had not shown favour to special relativity and did not mind the resulting restrictions on it. In contrast Nordström and Mie felt that they could not abandon a global constant velocity of light and chose $m=f(\Phi)$. Material relevant to the development of the gravitational theories which goes beyond what is given below, can be found in Abraham (1914*a*), Whittaker (1953), Harvey (1965*a*), Guth (1970), Hoffmann (1971), and Mehra (1974).

7.2. Gravitational theories: 1911–1915

Einstein had already abandoned the constancy of the velocity of light in 1907 when he had considered the implications of his new principle of equivalence. Using a different argument based on this principle he obtained the same expression for the velocity of light in 1911. In a gravitational potential Φ Einstein (1911) found that the velocity c was given by

$$c(\Phi) = c_0\left(1+\frac{\Phi}{c_0^2}\right) \tag{7.1}$$

where $c = c_0$ when $\Phi = 0$.

In 1912 Abraham published his first theory of gravitation (Abraham 1912*b*, *c*) in which he obtained

$$c(\Phi) = c_0\left(1+\frac{2\Phi}{c_0^2}\right)^{\frac{1}{2}} \tag{7.2}$$

which agrees with Einstein (equation (7.1)) to the first order of approximation. Soon after Einstein himself (Einstein 1912*a*) put forward a gravitational theory in which

$$\Delta c = \left(\frac{\partial^2}{\partial x^2}+\frac{\partial^2}{\partial y^2}+\frac{\partial^2}{\partial z^2}\right)c = kc\rho \tag{7.3}$$

where ρ is the density of matter and k is a universal (gravitational) constant. The gravitational force on a mass m was given by

$$F = -m \operatorname{grad} c. \tag{7.4}$$

Since the equations of this theory implied that action and reaction were not equal, Einstein (1912*b*) soon modified equation (7.3) to the form

$$c\Delta c - \tfrac{1}{2}(\operatorname{grad} c)^2 = kc^2\sigma \tag{7.5}$$

where σ is the density of matter and the energy density of other fields.

In this same year Abraham put forward a second theory at the International Congress of Mathematicians in Cambridge in August 1912 (Abraham 1912*d*). Abraham gave the following expression for the gravitational force k:

$$k = \frac{A}{r^2} - \frac{B}{r^3} \tag{7.6}$$

where $B/A = \gamma m'/2c^2$, γ being the gravitational constant and m' the mass of the central body.

This first stage in the development of a new gravitational theory was dominated by Einstein and Abraham who both put forward new theories and criticized at length the work of the other. In the second stage the main rival of Einstein was Gunnar Nordström, who proposed two gravitational theories in 1912 and 1913. Rather than make the velocity of light a function of the gravitational potential as Einstein and Abraham had done, Nordström (1912) chose to retain the unrestricted validity of special relativity, a move which required him to put

$$m = m_0 \exp(\Phi/c^2). \tag{7.7}$$

Later Nordström (1913) modified his theory: among other modifications the gravitational constant was now also potential dependent, being given by

$$g(\Phi) = \frac{g(\Phi_0)}{1 + (g(\Phi_0)/c^2)(\Phi - \Phi_0)}. \tag{7.8}$$

In 1913 Gustav Mie put forward a theory of matter which, though not important as a theory of gravitation, is generally considered to be influential in that it tried to encompass all physical processes in terms of a single world function by electric and magnetic quantities (Mie 1913).

Einstein also published an important paper in 1913. Early in 1911 he had gone to Prague to occupy a chair in physics, but had returned in 1912 to take up a similar position in Zurich. Here he collaborated with an old friend, the mathematician Marcel Grossmann, and together in 1913 they published an *Outline of a generalised theory of relativity* (Einstein and Grossmann 1913). This *Entwurf* paper was split into two parts, the first *Physical Part* being written by Einstein and the second *Mathematical Part* by Grossmann, and may be seen as the basis of the mature general theory of relativity. Einstein discussed the *Entwurf* theory at the 85th Congress of Natural Scientists at Vienna in December 1913 (Einstein 1913). Early in 1914 he moved to Berlin where more papers appeared, including a long *Formal Basis of the General Theory of Relativity* in November 1914 (Einstein 1914*a*). Finally late in 1915 the theoretical problems were overcome, and in four sessions of the Prussian Academy of Sciences in Berlin, the meetings of November 4th, 11th, 18th, and 25th, 1915, Einstein announced what we now know as the general theory of relativity (Einstein 1915). On November 18th he gave his explanation of the perihelion motion of Mercury and on November 25th the final field equations were given.

Because we are concerned with the empirical consequences of the preceding theories, and in particular with their predictions of perihelion motion (which will be given below), their theoretical nature has not been discussed. Nevertheless as far as Einstein was concerned there were at the end of 1913 (and until the end of 1915) only two theories worth discussing, those of himself with the initial collaboration of Grossmann and the second theory of Nordström which early in 1914 he and Fokker discussed in terms of the tensor calculus (Einstein and Fokker 1914). The two theories fall naturally into two classes, the gravitational potential in Nordström's theory being a scalar quantity and in the Einstein–Grossmann theory a tensor quantity (in its guise as the ten $g_{\mu\nu}$). As we shall see it was for a time impossible to decide between the two on purely empirical grounds, but Einstein had already argued strongly if not effectively against scalar theories.

In 1912 Einstein had given an argument against the first theory of Nordström, written in a letter to Nordström and cited at the end of the latter's first paper (Nordström (1912), p. 1129). According to Einstein the argument was first used against an early scalar theory of his own. He showed that in a gravitational field a rotating body would fall less quickly than a non-rotating one, a result in conflict with the principle of equivalence which stated roughly that in this situation, all bodies would fall with the same acceleration. The mere existence of this conflict was enough for Einstein, but Nordström felt that it was 'too insignificant to contradict experience'. We now know that a similar conflict exists in general relativity. In the 1960s, with an increase in interest in scalar theories, the argument was revived by Harvey (1964, 1965*b*) who showed that for a dumbell of length $2L$ rotating with angular velocity ω the difference in a landing time t would be to first order $(\omega^2 L^2/2c^2)t$ which is quite insignificant. However, Sexl (1967, pp. 305–7) has shown this argument to be incorrect.

A second argument which Einstein used against Nordström was concerned with Mach's principle. In his Vienna lecture of December 1912, Einstein said of Nordström's second theory:

> The only unsatisfactory thing is the fact that according to this theory the inertia of a body seems to be affected, although not produced, by other bodies. Because, according to this theory, the inertia of a body increases the farther the other bodies are removed from it. (Einstein 1913, p. 1254; transl. by Mehra 1974)

Einstein thought that he had shown that on his own theory:

> The inertia of a body must increase when ponderable masses are piled up in its neighbourhood. (Einstein and Grossmann, p. 228; transl. by Guth 1970)

This is the opposite of the effect in Nordström's theory. Guth (1970, pp. 190–3) has traced the path of the Einstein effect through Einstein's career and up to the early 1960s when Dicke noticed that this effect is non-invariant and only valid for a particular co-ordinate system. The effect does exist in Nordström's first theory and in the recent Brans–Dicke theory.

Einstein gave a further two arguments against scalar theories in a separate section of his *Entwurf* paper (Einstein and Grossmann (1913) § 7, pp. 242–4). At this time he still had no empirical arguments against them. He recognized, and admitted in the first sentence of this section, that his own theory appeared to be more complicated; Nordström's theory for instance contained only one gravitational potential as opposed to ten in Einstein's theory. In the scalar theory there was no coupling between electromagnetic and gravitational fields, which meant that the velocity of light was not altered near a massive body and that there would be no deflection of starlight in the Sun's gravitational field. Einstein considered that a box containing black-body radiation could be constructed which would lead to unacceptable consequences on a scalar theory but not on a tensor theory. This argument was withdrawn in a note inserted directly after Grossmann's mathematical part of the same paper (Einstein and Grossmann (1913) p. 261). Einstein had omitted to consider the variation of length in a gravitational field and the resulting stresses that would have been introduced into the system. Nordström (1913, p. 544) gave a detailed reply to Einstein and defended his theory.

Einstein concluded his *Entwurf* paper by stating that to him the most effective argument against scalar theories was that relativity existed with respect to a much greater transformation group than the linear orthogonal group (the Lorentz group) admitted in those theories. Nevertheless he qualified this by adding that he had not yet found the most general transformation group belonging to his equations of gravitation.

These four arguments may have been sufficient to keep Einstein on his chosen path to a generalized theory of relativity, but they did not form a strong list. The empirical arguments, as we shall now see, were also not clearly for or against any particular theory.

7.3. Perihelion predictions: 1911–1915

We may start the comparison of empirical predictions by looking at the consequences for perihelion motions of each of the gravitational theories mentioned at the beginning of this chapter. In view of the importance attached to the Mercury anomaly it is perhaps surprising to find that the respective predictions were usually given little prominence, often being given in subsidiary papers or sometimes not at all. A reason for this is suggested below.

The empirical consequences of Abraham's first theory were worked out by Pavanini (1912, 1913). In the second paper the secular motion of the perihelion of Mercury was calculated to be $+14''.52$, i.e. about one-third of the general relativistic value. Abraham cited his quantity in his *Jahrbuch* review article of 1914 (Abraham 1914*a*, § IIB) but gave no values for other theories. Nowhere was the Newcomb anomaly of $+42''$ mentioned.

The 1912 theory of Einstein was not worked out in terms of its empirical consequences. Whitrow and Morduch (1960, 1965) have calculated that this theory gives two-thirds of the general relativistic value.

Abraham's second theory was discussed by Caldonazzo (1913) but no applications were given. We can use Newton's formula which states that for a force of the form

$$F = br^{m-3} - cr^{n-3}$$

the angle between successive perihelia is $2\pi\{(b-c)/(mb-nc)\}$. If we apply this to Abraham's force law (equation (7.6)) and note that $B/A = 10^{-8}$ for the Sun (Abraham 1912d, p. 797), can see that the angle between successive perihelia is $2\pi\,(1\text{–}10^{-8})^{\frac{1}{2}}$ which to first order involves a regression of perihelion of $\pi \times 10^{-8}$ rad per orbit or about $3''$ per century.

Nordström did not give the empirical consequences of his first theory. Behacker (1913) gave the equation of a planetary orbit but did not carry this further. He remarked:

> The gravitational theory of Nordström reproduces free fall in empty space and the motion of a planet in the gravitational field of the Sun in a manner fully corresponding with experience. (Behacker 1913, p. 992)

Whitrow and Morduch (1960, p. 792) have shown that this theory gives a regression of perihelion equal in magnitude to one-third of the general relativistic advance, which does not at all correspond to experience unless some other explanation is given for the Mercury anomaly.

Nordström (1914) himself discussed planetary motion on his second theory. He obtained for the perihelion motion for each orbit:

$$\delta = \frac{g_0^2 g_1^2 (1-q_1)^2 m^2}{c^2 f^2 (4\pi)^2} \tag{7.10}$$

where $g_0 = c^2/\Phi_0'$, for which the gravitational potential Φ' is given by $\Phi' = \Phi_0' - M/4\pi r$ for a central body of mass M. The quantity q_1 is the value of q at the planet's aphelion distance, where $q = v/c$ and v is the velocity of the planet. According to Nordström:

> g_0 and g_1 have very nearly the same value; in our part of the Universe $g_0 = 9.15 \times 10^{-4}$ $\mathrm{cm}^{3/2}\,\mathrm{sec}^{-1}\,\mathrm{g}^{-\frac{1}{2}}$. For the motion of the Earth about the Sun one obtains for δ a value of approximately $\delta = 10^{-8}$. The angle $360°/\sqrt{(1+\delta)}$ differs from $360°$ by only about 0.0065 seconds of arc.
>
> This rotation of the ellipse of the Earth's orbit every year is therefore very small in proportion to the astronomical perturbations. (Nordström 1914, p. 1108)

The value for Mercury is surprisingly not given. The regression for the Earth's perihelion is about $0''.65$ per century on this theory. Only the function f is a function of planetary properties (q_1 only to a small extent); it is the constant in the area law $r^2 \mathrm{d}\theta/\mathrm{d}t = f$ so that $f^2 = GMa(1-e^2)$. I have evaluated δ for Mercury and found a regression in the perihelion of about $6''.9$ per century.

This is confirmed in recent work, though one has to take care to specify which of Nordström's theories is being discussed. Schild (1962), for example, ended his paper on gravitational theories of the Whitehead type with a comparison of general relativity and Nordström's theory. The reference indicates that it is the second theory that is being discussed. Schild obtained a perihelion rotation of '1/6 amount predicted by general relativity and in the opposite direction'. He also referred to the 1953 theory of D.E. Littlewood, who obtained the same field equations as those in Nordström's second theory (Schild 1962, p. 113). The perihelion shift on Littlewood's theory was calculated by Gürsey (1953), but he made an error which was point out by Das (1957) after Pirani (1955) had obtained the correct result, i.e. a regression of one-sixth of the general relativistic value. Whitrow and Morduch (1960, p. 792) confirm this result and point out the empirical identity of this theory with that of O. Bergmann, but what they call Nordström's theory is his first theory.

Mie's theory would seem to give no perihelion motion, on the grounds that it gave the Newtonian law of attraction (Mie 1913).

7.4. Einstein, perihelion predictions, and Freundlich

Einstein never mentioned the perihelion advance of Mercury in published papers until the 18th November 1915, when he announced that the whole of the anomalous advance was accounted for by general relativity. He limited himself in the *Entwurf* theory to showing that Newtonian motion would be obtained in the limit. That a perihelion advance was obtained in this theory was first shown in a paper read in Amsterdam at the end of December 1914 by Droste (1915). He derived an expression for the advance using calculations of the Einstein–Grossmann field made by Lorentz, and added

> As Prof. de Sitter has calculated from the equations of motion determined by Prof. Lorentz, it amounts for Mercurius to 18″ per century, the observed motion being 44″. (Droste 1915, p. 1010)

It is a moot point whether Einstein drew the empirical consequence of his theories for himself as they developed. That most of his gravitation papers of this time give red shift and light deflection predictions but never give or even mention perihelion results suggests that he did not pursue the perihelion question. He did know of the *Entwurf* prediction, for he wrote to Arnold Sommerfeld on the 28 November 1915 that his previous field equations were 'entirely untenable' and that one of the reasons that he had known this was that the motion of the perihelion of Mercury was '18″ instead of 45″ [*sic*] per century' (Hermann 1966, p. 32). Though the use of 45″ suggests a certain vagueness, there seems no reason why he should not have got this value from Droste or de Sitter.

That his final theory gave the full perihelion advance certainly gave Einstein great pleasure. In December 1915 he wrote to Michele Besso that 'what pleases me most is the agreement with the perihelion motion of Mercury' (Speziali 1972), and to Sommerfeld, 'The result of the perihelion motion of Mercury fills me with great satisfaction' (Hermann 1966, p. 37). No wonder he was pleased, for his new theory was both theoretically successful (he previously had had to abandon general covariance) and empirically successful. However, what was the nature of the latter success, and why could he claim credit for explaining an anomaly that had been so little mentioned in the relevant papers of the preceding years? We cannot pretend that he did not know of the anomaly in Mercury's perihelion or that he did not think his work relevant, for already in 1907 Einstein had written in a letter that he was

> busy on a relativistic theory of the gravitational law with which I hope to account for the still unexplained secular change of the perihelion motion of Mercury. (Seelig 1956, p. 76)

Nor was Einstein alone in this public neglect of Mercury's perihelion. Of the seven theories of gravitation that have been considered (two each from Abraham, Einstein, and Nordström, and the theory of Mie) before general relativity, none of the main papers of the authors mentioned this anomaly, and only Pavanini's paper on Abraham's first theory and Droste's paper on the *Entwurf* theory gave evaluations for Mercury, and even so Pavanini did not mention that there was an anomaly. Discussion in these papers was concerned with general principles and whether or not they were satisfied. We have considered Einstein's theoretical arguments against the scalar theories. Given in a review paper of 1914, the following remarks against the theories of Abraham and Mie were not untypical of Einstein:

> Theories of gravitation have been established by Abraham and by Mie. The theory of Abraham is in contradiction with the principle of relativity, that of Mie with the condition of equality of inertial and gravitational mass of isolated systems. On this last theory, when one heats a body the inertial mass of this body increases as the energy increases, but not the gravitational mass; for a gas, this will diminish even when the temperature is increased. (Einstein 1914b, p. 145)

Einstein at this point added a note in which he recognized that the temperature effect on Mie's theory was too small to observe, but that he wanted the equality to hold in principle. Yet two other observational effects were being mentioned. In almost every relevant paper of Einstein one finds it stated that the theory gave a red shift to light omitted in a gravitational field and that a ray of light passing close to the Sun would be deflected by $0''.83$, and in the main paper on his second theory by Einstein's chief rival between 1913 and 1915, Nordström, one also finds it stated that the theory gave a red-shift of the same value as that of Einstein (Nordström 1913, p. 549). Since Nordström retained the constancy of the velocity of light his theory gives no light deflection in a gravitational field. As far as perihelion motions were concerned, we have seen that Nordström calculated the motion for the Earth alone on his second theory and that he had commented that the effect was very small compared with the astronomical perturbations. He concluded:

> . . . it can be said that the laws derived for free fall and planetary motion agree very well with experience. The same would be true, as Hr. Behacker has shown, if one based the calculation on my original gravitational theory. A comparison between the laws of motion contained in both these theories certainly shows that the laws in the new theory assume a simpler and more harmonious form. (Nordström 1914, p. 1109)

Assuming that Nordström calculated the perihelion motion for Mercury, an easy step given the formula (7.10), why did he think that the result, a $7''$ regression per century, agreed with experience? He was correct when he stated that the effect was small compared with the perturbations of the other planets—these in the case of Mercury for instance total some $500''$ but this is irrelevant to the point in question. One must take it as most unlikely that this was the reason why he thought his theory agreed with experience, since it implies that Nordström was ignorant of 60 years of discussion concerning the anomalous perihelion of Mercury. However, the advance to be explained on Newtonian theory was $42''$, so that the advance to be explained on the

Nordström theory was 49″ which is not really much more than before. If the 42″ was not anomalous on Newtonian theory, neither was the 49″ on Nordström's theory. I think that the reason for Nordström's attitude and the general neglect of the perihelion of Mercury as an anomaly to be explained by any new gravitational theory was that Seeliger's hypothesis was being taken very seriously. Seeliger, as we saw in Chapter 4, had in 1906 explained the 42″ by attributing it (and some other planetary motion anomalies) to diffuse matter surrounding the Sun, part of it being the matter causing the zodiacal light. On this view the 42″ was not an anomaly for Newtonian theory. If Nordström felt that his extra 7″ could also be accounted for in this way he was not alone, for in 1918 Harold Jeffreys thought that he could explain an extra 7″.4 regress on a theory of Silberstein by a slightly greater density of matter than Seeliger had used.

Einstein also did not appear to be worried by the Nordström regress in perihelia. His 1914 review from which an extract has already been given continued by referring to Nordström:

By contrast [to Abraham's and Mie's theories], a theory of gravitation recently established by Nordström agrees as well with the principle of relativity as with the condition of the gravitation of energy of isolated systems, with a restriction indicated further on. The declaration to the contrary of Abraham, in the study that appeared in this review, is not just. I believe besides that one cannot find in experience a valid argument against the theory of Nordström. (Einstein 1914b, p. 145)

However, he did think that such an argument could eventually be found. At the Vienna conference in December 1913 Einstein discussed the *Entwurf* theory and Nordström's second theory. Nordström had met Einstein in Zurich during 1913, and the paper in which he proposed his second theory was dated July 1913, Zurich. The paper which gave the perihelion regress was dated January 1914 (Helsinki), so one cannot tell whether Einstein knew of this effect the previous month. Einstein's conclusion was that the only way of deciding empirically between his approach and that of Nordström was by measuring the deflection of starlight during an eclipse since Nordström's scalar theory gave no such deflection. He added that 'hopefully the solar eclipse as early as that of 1914 will bring about the important decision (Einstein 1913, p. 1262). He repeated the hope in his farewell lecture given in Zurich on February 9th 1914 when he moved to Berlin:

A decision between both theories through experiment is possible in so far as according to the Einstein–Grossmann theory, but not according to the Nordström theory, the gravitational field must cause a bending of light rays. Since the only gravitational field which would produce a bending of rays accessible to observation is that of the Sun, careful preparations are being made for the solar eclipse taking place in August 1914: it should be ascertained through photographic exposures of the fixed stars near the Sun whether this bending of rays is in fact present or not. (Einstein 1914c, p. 6)

It was Einstein's astronomical contact Erwin Freundlich who set out to measure the starlight deflection at the 1914 eclipse.† It was not the first such expedition; that was made by the Cordoba Observatory at the 1912 eclipse (Einstein had given a quantitative prediction of the effect in 1911) but it was stopped by bad weather (von Kluber 1960, p. 47). Freundlich chose to make his observations in the Crimea and the party reached there, but unfortunately the First World War started and he was captured by the Russians, returning to Berlin in September 1914 after an exchange of prisoners. It now appeared that there was no immediate hope of distinguishing between the theories of Nordström and Einstein as far as experimental success was concerned. The light deflection test could not be performed under conditions of war, and proposals to use existing photographs or daylight photography or to use Jupiter as the deflecting body were impractical. In addition both theories gave the same red shift, so that the work on the solar red shift being done at the time could not have been used to differentiate between them even if anyone had been certain that the observed red shifts were gravitational in origin.

The only other difference could have been in the perihelion predictions. By the beginning of 1915 it was known that the Einstein—Grossmann theory gave an 18″ advance for Mercury (Droste's result). If one accepted that Newcomb's 42″ was indeed anomalous, then Einstein's value was clearly preferable, for his remaining anomaly was 24″ compared with Nordström's 49″. Since it was felt by both Einstein and Nordström that no empirical argument existed beyond the light deflection predictions, one can only assume that the perihelion motion of Mercury was not considered to be anomalous and that the prevailing hypothesis explaining it, Seeliger's hypothesis, was valid. Freundlich, in his desire to test Einstein's ideas and having failed to carry out what appeared to be the only distinguishing experiment, would at the end of 1914 have turned naturally to the other prediction that was different in the two theories, the motion of planetary perihelia. Since it was Seeliger's hypothesis that stopped the Mercury anomaly from serving as evidence either for or against the theories, we should not be surprised that it was to this that Freundlich next turned his attention. On February 27th 1915 he sent a paper to the *Astronomische Nachrichten* in which he showed that Seeliger's hypothesis could not be used to account for any of the advance in Mercury's perihelion (Freundlich 1915).

It has been remarked that the timing of Freundlich's paper seemed odd in that Seeliger had published his hypothesis in 1906, eight years before, and it has been suggested that early in 1915 Einstein had found his final field equations and had obtained the full perihelion shift,‡ and thus required that Seeliger's hypothesis was shown to be untenable. This thought conflicts with Lanczos's statement that the final field equations could not have been

† For an account of Freundlich's relationship with Einstein see Pyenson (1976).

‡ A similar suggestion is to be found in Pyenson (1976).

obtained before the autumn of 1915 (Lanczos 1974). The reconstruction given above seems to give a better explanation of why Freundlich's work was done early in 1915. It also explains why so little emphasis was given to the perihelion motion predictions of the theories that we have been discussing.

This explanation is not watertight. Einstein himself, in a letter to Sommerfeld written on 2nd February 1916, played down Freundlich's attack on Seeliger's hypothesis:

> Freundlich was the only colleague who effectively supported me in my efforts in the area of general relativity. He has devoted years of thought and also of work to the problem so far as this was possible in addition to the trying and dull service to the observatory.
>
> Freundlich has in addition a second merit. I do not refer to the refutation of Seeliger's theory of the perihelion motion of Mercury, for this action is perhaps to be described as forcing an open door. But Freundlich has shown that modern astronomical resources suffice for the purpose of proving the light deflection at Jupiter, which I would not have considered possible, although I already thought about the case some years ago. Contact with astronomy is the very thing I lack. (See Hermann 1968, p. 380)

The deflection of starlight by Jupiter according to general relativity is only 0″.02, and Freundlich's hope of measuring this proved to be unfounded. Perhaps one should not take Einstein's remarks too seriously. After all, even if Seeliger's hypothesis was an open door until then, and even after, it was taken as being sufficient to account for the Mercury anomaly. Also, as we have seen in Chapter 4, it accounted for the motion of the nodes of Venus, which was anomalous for general relativity for a long time.

One could also cite the paper of Droste, in which he gave the 18″ *Entwurf* theory advance, as referring to the observed 44″. This paper was delivered before Freundlich's paper, and his reference to the 'observed motion being 44″' indicates that he thought that the amount of requiring explanation was the full 44″. However, this was consistent with the attitude towards Seeliger's hypothesis shown by de Sitter who had calculated the 18″. We have seen in Chapter 4 how de Sitter viewed Seeliger's hypothesis as merely determining the amount of matter surrounding the Sun from the excessive motion of Mercury's perihelion. The hypothesis could account for all the Newcomb anomaly, but if a 'better' theory itself predicted a perihelion advance then this just showed that there was less circumsolar matter than had previously been thought. In his 1916 papers introducing general relativity to England, de Sitter did not describe Seeliger's hypothesis as untenable but as 'superfluous'. In this situation it is not surprising that the perihelion advance of Mercury was not taken as discriminating evidence for the new gravitational theories, and it is not surprising that Freundlich, in his capacity as Einstein's empirical arm, should have decided to attack Seeliger's hypothesis. Having done so, there was at last empirical evidence to show the superiority of Einstein's theory over that of Nordström, and that was its limited function in early 1915 for

Einstein's preference for judging theories on *a priori* grounds would have stopped him from returning to his 1912 theory which had given a 28″ advance to Mercury's perihelion. The empirical way was now clear for Einstein to finish his theory of gravitation, which he did, after months of frustration, in the autumn of 1915.

I propose to treat the general relativistic prediction of the precession of planetary orbits by considering four components of the theory. In developing his theory Einstein used the principles of equivalence and covariance. He used a non-Euclidean geometry to describe space-time intervals. Finally he required that special relativity was valid in the absence of a gravitational potential and locally in such a potential. These four components will now be considered with regard to their effect on the motion of planetary perihelia.

7.5. Special relativity and the perihelion advance

Special relativity was introducd in 1905 as a theory of electrodynamics, and one of its consequences was a variation of mass with velocity. It was not unique in this, for the electron theory of Lorentz gave the same dependence on velocity and similar relations had been put forward by Abraham, Langevin, and Bucherer. The experiments of Kaufmann favoured the latter three, but the 1908 experiments of Bucherer showed the Lorentz–Einstein formula to be more correct and his results were upheld by later workers. The relativistic formula applied to the whole mass, whereas before mass variation with velocity was thought to be confined to the electromagnetic component of the mass with a 'proper' mass remaining constant. The effect of a velocity variation in electromagnetic mass had been considered by astronomers as early as 1902, as George Darwin related in an introductory lecture on dynamical astronomy:

> The newer theories of electricity with which the name of Prof. J.J. Thomson is associated indicate the possibility that mass is merely an electrodynamic phenomenon. This view will perhaps necessitate a revision of our accepted laws of dynamics. At any rate it will be singular if we shall have to regard electrodynamics as the fundamental science, and subsequently descend from it to the ordinary laws of motion. How much these notions are in the air is shown by the fact that at a congress of astronomers, held in 1902 in Göttingen, the greater part of one day's discussion was devoted to the astronomical results which would follow from the new theory of electrons. (Darwin 1916, p. 10)

Darwin must be referring to the meeting of the *Astronomische Gesellschaft* at Göttingen from 4th to 7th August 1902 at which he read a paper. The relevant discussion seems to have been the following.

> In the afternoon [5th August] at 3 o'clock there took place in the Chemical Institute of the University a lecture combined with demonstrations of cathode- and Bequerel-rays by Herren Prof. Riecke and Dr. Kaufmann, which was followed by a lively debate. Furthermore Herr Lebedev spoke about the physical causes of deviations from the Newtonian law of gravitation. (*Vierteljahrsschr. astr. Ges.*, 1902)

Lebedev's paper was mostly concerned with cometary theory and the motions of cometary tails, so it was the Riecke–Kaufmann lecture and the ensuing 'lively debate' that Darwin was recalling.

Before this, in 1900, Wien had discussed the effect of velocity-dependent masses (Wien 1900, 1901). He referred to a paper of Lorentz of 1900 in which various electrodynamic theories of gravity were considered (Lorentz 1900). This paper has been considered in Chapter 6; we may merely note that in it Lorentz had to take account of the Sun's motion through the ether. Wien found that the effect of an electromagnetic mass—using Searle's expression for the energy of a charged ellipsoid in motion—was formally the same as the use of a conventional mass in Weber's law of attraction, with a numerical factor of unity replaced by 15/16. He added

> It is well known that Weber's law has been used with certain success in the theory of Mercury's motion. (Wien 1901, p. 509)

Thus one would expect a similar advance to that obtained on Weber's law, i.e. about 7″ per century. Wilkens (1904) discussed the effect of electromagnetic masses in terms of secular changes to the parameters of the planetary orbits at the meeting of the *Astronomische Gesellschaft* at Lund. He obtained for the secular advance in Mercury's perihelion a value of 3″.5 per century and concluded that 'the views of the new electrodynamics are compatible with astronomical observations' (Wilkens 1904, p. 212). He republished his results in 1906 (Wilkens 1906), apparently in reply (or rather to claim precedence) to a paper by Wacker published slightly earlier that year (Wacker 1906). When ether theories with their absolute velocities were used, the velocity of the Sun relative to the ether played a role. Assuming that the Sun was at rest in the ether, Wacker obtained as the centennial advance in the perihelion of Mercury 5″.8 on Abraham's hypothesis of the mass variation and 7″.2 on Lorentz's hypothesis. Taking Abraham's hypothesis of rigid electrons (in 1906 it seemed from Kaufmann's experiments that Abraham's hypothesis was better confirmed than that of Lorentz), this quantity was recalculated on the assumption that the ether was fixed in the system of the fixed stars. The Sun thus moved relative to the ether, but this motion depended on what one took to be the apex of the solar motion. If it was $A=270°$, $D=36°$, the perihelion motion was 6″.5, but if the apex was $A=270°$, $D=0°$, the motion was 5″.0. Wacker noted that these values were small compared with the observed 40″, and he asked whether the absolute velocity of the solar system might be determined so as to represent all the planetary motions.

Wacker's results were also obtained by Poincaré in lectures given to students in 1906–7 (Poincaré 1953, see Chaps. 11 and 12 and p. 236). He gave a general formula for the perihelion advance which reduces to about 5″.6 for Abraham's hypothesis and 7″ for that of Lorentz.

However, this trend soon stopped, with only Wacker's 1909 Tübingen thesis under R. Gans appearing later (Wacker 1909). The new line was to find a law of gravitation that was Lorentz invariant and that reduced to the Newtonian law for bodies at rest. The first to do this was Poincaré (1906, pp. 129–176; see §9, pp. 538–50)† but he found that, having obtained the general form of the solution, there were any number of actual solutions. Simplicity restricted him to an examination of two hypotheses. One contained a velocity of propagation of gravitation which was greater than that of light, but this in certain cases required an advance effect (the gravitational action depending on the position of a body that it had not yet attained) which Poincaré was not prepared to countenance. He preferred the second in which the velocity of propagation was equal to that of light. This appeared to suffer from Laplace's argument against such velocities (for which see Chapter 6), which the other hypothesis did not, but

> ... Laplace had examined the hypothesis of the finite velocity of propagation *ceteris non mutandis*; here, on the other hand, this hypothesis is complicated by many others, and it could occur that among them there was a more or less perfect compensation, as those giving us applications of the Lorentz transformation have already given us so many examples. (Poincaré 1906, p. 544)

Poincaré gave no astronomical consequences of the law, but since the force contains, in addition to a component along the line of centres, a component parallel to the attracting body, it is clear that a perihelion motion would result. In a paper of 1908 Poincaré devoted a section to the astronomical consequences, and found that for a fixed attracting centre his hypothesis was observationally indistinguishable from the ordinary Newtonian law coupled with Lorentz's deformable electrons (Poincaré 1908, pp. 386–402, 577–9, 580–1). Both gave 7″ for the centennial advance in Mercury's perihelion as opposed to 5″.6 for Abraham's hypothesis and 14″ for Weber's law. I have argued in Chapter 6 that this 14″ is incorrect (it was originally due to Tisserand) and should be 7″. Poincaré's values thus confirm the work of Wien and Wacker. Poincaré concluded

> In summary, the only sensible effect on astronomical observations will be a movement of the perihelion of Mercury, in the same sense as that which has been observed without being explained, but notably weaker.
>
> That may not be regarded as an argument in favour of the new dynamics, since it would always be necessary to look for another explanation for the greater part of the Mercury anomaly; but that again may not in the least be regarded as an argument against it. (Poincaré 1908, p. 581)

In his 1906–7 lectures Poincaré expressed preference for a circumsolar ring of matter to explain Mercury's perihelion; the density of the ring was chosen

† Poincaré (1953, pp. 149, 265) where he stated in conclusion:
At the moment we have no satisfactory explanation of attraction. On the other hand we have no serious reason to modify the Newtonian law. The most grave discordance is the advance of the perihelion of Mercury. But it is probable (*vraisemblable*) that this anomaly is due to the existence of a ring, such as we have seen in Chapter IV.

to suit the required numerical value of the anomaly.† As has been emphasized above, the adoption of this hypothesis made it impossible to use the perihelion anomaly as a test of a gravitational theory unless independent assessments were made of the density of the disturbing matter.

Minkowski (1908) gave a Lorentz-invariant form to the Newtonian law, but his law produced no observable difference to the conventional form. Lorentz himself produced an expression similar to that of Poincaré, resulting in a 6″.69 advance per century for Mercury's perihelion (Lorentz 1910, pp. 1239–40).‡ He noted the difficulties of deciding on this with Seeliger's hypothesis still alive, but hoped to measure the motion of the solar system through the ether by the use of a method first suggested by Maxwell, i.e. by determining whether the times of the eclipses of Jupiter's satellites were a function of the position of Jupiter with respect to the Earth. Fixing this velocity would of course then fix the perihelion motion on the ether theory.

de Sitter (1911) gave his own treatment of these results. He criticized the work of Wilkens, discussed Minkowski's law and also the law of Poincaré and Lorentz. He noted that the only observable consequence of adopting a relativistic law was a 7″.15 advance per century in Mercury's perihelion and that Seeliger's hypothesis could explain the whole anomaly. de Sitter also pointed out that one could obtain any multiple of 7″.15 by a suitably formulated Lorentz-invariant law. If, instead of the invariant dimensionless quantity C, the invariant quantity C^n were used in expressing the forces, a perihelion advance of 7″.15n was obtained. Obviously one could obtain the Newcomb anomaly with $n=6$. This was the equivalent of Poincaré's realization that there were any number of satisfactory laws; for him, simplicity dictated the choice of $n=1$. More recently Whitrow and Morduch (1960, p. 792) have given a similar treatment of 'Poincaré's law.

7.6. Non-Euclidean geometry and the perihelion advance

Talk of curved space in general relativity is an elliptical way of talking of curved space-time. The general expression for the line element ds is

$$ds^2 = g_{\mu\nu}\, dx_\mu\, dx_\nu \qquad (7.11)$$

in which the Einstein summation convention is used and μ and ν range from 1 to 4 for which $\{x_1,x_2,x_3,x_4\}=\{x,y,z,ict\}$. The Euclidean or flat space-time line interval is given by

$$ds^2 = dx^2+dy^2+dz^2-c^2\, dt^2. \qquad (7.12)$$

Einstein, however, made the ten quantities $g_{\mu\nu}$ (ten since $g_{\mu\nu}=g_{\nu\mu}$) functions of the gravitating matter in the region under consideration so that the

† This was published in abstract form in 1905.

‡ Lorentz thought Seeliger's hypothesis plausible, but attributed the discovery of the Mercury anomaly to Laplace!

Newtonian potential enters eqn. (7.11) via the $g_{\mu\nu}$ and except in the absence of matter the line element ds is no longer flat or Euclidean.

The notion of space-time was introduced by Minkowski in 1908, but before this curved space had been applied to the planetary motions. Much of the work was orientated towards pure mathematics and few papers discussed explicit observational consequences. A central question was what to adopt as the law of attraction. In a spherical space of radius of curvature R the distance $2q$ between two points is replaced by the distance $2r$ as shown in Figure 7.1.

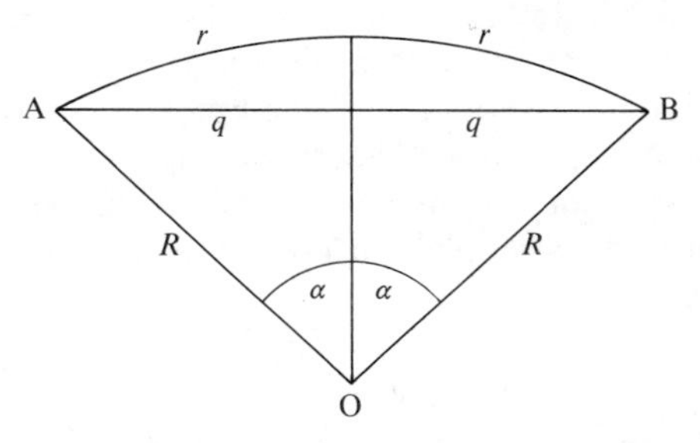

FIG. 7.1. Distance in flat and curved space.

We can see that $q = R \sin \alpha$ and that $r = R\alpha$ where α is expressed in radians. Hence we obtain

$$q = R \sin(r/R). \tag{7.13}$$

A heuristic argument for an inverse square law is that the flux measured on a given surface element on a spherical surface of radius r decreases with the inverse square of r since the surface area of a sphere is $4\pi r^2$. The surface area of such a sphere in our curved space is $4\pi R^2 \sin^2(r/R)$, so that on this argument the general law of gravitation is, using conventional symbols,

$$F_R = \frac{GMm}{R^2 \sin^2(r/R)}. \tag{7.14}$$

This is derivable from a potential function

$$U_R = \frac{GMm}{R \tan \alpha} \tag{7.15}$$

since

$$\begin{aligned} F_R &= -\frac{\partial U}{\partial r} = -\frac{\partial U}{\partial \alpha}\frac{\partial \alpha}{\partial r} \\ &= -\frac{GMm}{R}\left(\frac{-1}{\tan^2 \alpha}\right) \sec^2 \alpha \, \frac{1}{R} \\ &= \frac{GMm}{R^2 \sin^2(r/R)}. \end{aligned}$$

The form given by equation (7.14) was used by Killing (1885) and the form given by equation (7.15) was used by Carl Neumann (1886). They both give the equivalents of Kepler's laws for a spherical space. A similar result was obtained by Liebmann (1902) using hyperbolic space. His attractive force was inversely proportional to $R^2 \sinh^2(r/R)$. Liebmann in fact followed the method of Bertrand, who in 1873 had discussed the problem of which distance-dependent forces gave closed orbits. It was this paper of Bertrand that Asaph Hall had used in 1894 when proposing Hall's hypothesis, which Simon Newcomb used in 1895 to obtain the advance in Mercury's perihelion, but Liebmann did not mention this application.

Killing and Neumann had obtained an elliptical orbit that did not precess by using the analogous form of the Newtonian law of attraction. Lense (1971) considered the question of what would happen if the Newtonian law was in fact universally true but was applied in a spherical elliptical space. In that case we continue to use

$$F = \frac{GMm}{r^2} \tag{7.16}$$

but we have to substitute for r according to equation (7.13) which gives us

$$F = \frac{GMm}{R^2\{\arcsin(q/R)\}^2} \approx \frac{GMm}{q^2} - \frac{GMm}{3R^2}. \tag{7.17}$$

The second term of equation (7.17) as a constant force added to the inverse square force produces a perihelion motion. Newton's theorem for the required effect of such forces tells us that the motion will be an advance of $2\pi/3R^2$ per revolution. Lense obtained for the advance per revolution

$$\Delta\varpi = \frac{\pi GMc}{3R^2\,\mathrm{h}^{\frac{3}{2}}}. \tag{7.18}$$

From stellar parallax considerations Schwarzschild had in 1900 given a lower bound for R, which gave a value of $10^{-9''}$ for the centennial advance of Mercury's perihelion from equation (7.18). Realizing the small magnitude of the effect Lense repeated his calculations using formulae of celestial mechanics for the changes caused by perturbing forces, this perturbing force being taken as the second term in equation (7.17). He found that the secular advance in the perihelion per unit time was $c/3R^2$, where c is the planetary velocity, which agrees with our 'Newtonian' value as one revolution takes $2\pi/c$ units of time.

In order to obtain a perihelion advance of the required amount one needs a much smaller value of R, i.e. a much greater curvature. Since this would be too great for the Universe, the curvature must be local. In naïve terms, Einstein made the curvature local and also variable by making it a function of the mass of massive bodies. Thus the curvature (in fact of space–time) was greatest nearer the Sun, and the planet most affected was the one nearest the Sun, i.e. Mercury.

The next problem was to determine the nature of the connection between matter and the curvature of space–time. The often cited early conceptions of Clifford (1876) that matter could be envisaged as local curvature of space, material particles being steep hillocks in a Euclidean plane, did not help. The suggestion of FitzGerald (1894, see Larmor 1902, p. 313) that 'Gravity is probably due to a change of structure [of the ether], produced by the presence of matter', was closer, but gave no clue to the nature of the change. What was needed was (a) a quantitative relation between matter and surrounding space and (b) an underlying theory which yields and explains (a). We shall see below how the principle of equivalence gives a clue to the nature of (a) when this is interpreted as space-time. The general relativistic version of (a), the Schwarzschild line element, is sufficient to describe the three classic predictions of general relativity. However, even this may be obtained by other theories, and one cannot step uniquely from (a) to general relativity. It was Einstein's field equations which supplied (b), from which at the end of 1915 he used approximate methods to obtain (a) and from which early in 1916 (and closely followed by Droste) Schwarzschild obtained exactly what we call the Schwarzschild line element. We shall not be concerned with the field equations.

Finally, one should perhaps note that since Lense's paper appeared in 1917 the discussion of the last few paragraphs has been rather ahistorical. However, its contents could have been given at any time after Schwarzschild's 1900 paper on the possible curvature of the Universe.

7.7. The principle of equivalence and the perihelion advance

The principle of equivalence was introduced by Einstein in 1907 and thereafter held a central position in the development of general relativity (Einstein (1907). It placed on a fundamental basis the long known fact of the equality of gravitational and inertial masses, an equality that had been experimentally established to high accuracy by Eötvös in 1890. Ignoring for the present criticisms and sharper formulations of the principle, it states that an observer cannot distinguish between the effects of being in a homogeneous gravitational field and in a uniformly accelerated frame of reference.

The strongest assertion of the power of this principle with respect to the perihelion of Mercury was made in 1945 by Kai-Chia Cheng, who published a method of calculating the perihelion advance directly from the principle of equivalence:

> It is known that first order phenomena described in the general theory of relativity can be deduced from the principle of equivalence. The present note serves as an illustration.
>
> The principle of equivalence demands that the length along the field direction be

deviated, to the first order of approximation, in the ratio: $\mathrm{d}r{:}(1+\lambda/r)\,\mathrm{d}r$, where λ is GM/c^2; and that the time, $\mathrm{d}t$, be modified into the expression $(1+\lambda/r)^{-1}\mathrm{d}t$. But the length normal to the field direction undergoes no change.

Hence the kinetic energy is expressible in the form $\frac{1}{2}(\beta^4\dot{r}^2+\beta^2\dot{\theta}^2r^2)$ and the Lagrangian is accordingly $\frac{1}{2}(\beta^4\dot{r}^2+\beta^2\dot{\theta}^2r^2)-V(r)$, where $V(r)=-GM/r$, $\beta=(1+\lambda/r)$. (Kai-Chia Cheng 1945)

Having set up the Lagrangian equations, Cheng proceeded to obtain from these, in a tortuous manner which was not given in detail but which I have checked, the equation

$$\frac{\mathrm{d}^2u}{\mathrm{d}\theta^2}+u=\frac{GM}{h^2}+3\lambda\,u^2 \tag{7.19}$$

where $u=1/r$ and $h=\beta^2r^2\dot{\theta}=$constant. Since this is also what is obtained in general relativity, the same perihelion motion is found.

A colleague of Cheng, Miss Su-Ching Kiang, published a note in 1946 in which she derived from eqn. (7.19) a value of the light deflection in the Sun's gravitational field of 2″.61, but by using an incorrect method. She concluded, however:

This result seems in better agreement with corrected astronomical data than Einstein's original value, which is only two thirds of the above value. (Su-Ching Kian 1946)

Now measurements of the light deflection were never in good agreement. Early ones gave fair support to general relativity but in 1929 an expedition led by Freundlich measured the deflection to be 2″.24 ± 0″.10, which is greater than the Einstein value of 1″.75. Then in 1936 a Russian measurement gave 2″.73 ± 0″.31, and it was probably this that Miss Kiang was referring to. Although the full report was not published until 1949, an abstract of a Russian paper giving a value of 2″.71 ± 0″.26 obtained in 1940 was published in the *Physikalische Berichte* of 1941.† Such a high value was not found again.

We can consider Cheng's assertion as follows. The flat space-time line element $\mathrm{d}s$ is given, in polar coordinates, by

$$\mathrm{d}s^2=\mathrm{d}r^2+r^2\mathrm{d}\theta^2+r^2\sin^2\theta\mathrm{d}\varphi^2-c^2\,\mathrm{d}t^2. \tag{7.20}$$

Then according to Cheng the principle of equivalence requires

$$\mathrm{d}s^2=\left(1+\frac{\lambda}{r}\right)^2\mathrm{d}r^2+r^2\mathrm{d}\theta^2+r^2\sin^2\theta\mathrm{d}\varphi^2-c^2\left(1+\frac{\lambda}{r}\right)^{-2}\mathrm{d}t^2$$

or to first order

$$\mathrm{d}s^2=\left(1+\frac{2\lambda}{r}\right)\mathrm{d}r^2+r^2\mathrm{d}\theta^2+r^2\sin^2\theta\mathrm{d}\varphi^2-c^2\left(1-\frac{2\lambda}{r}\right)\mathrm{d}t^2. \tag{7.21}$$

† See *Phys. Ber.* **22,** 934 (1941). This is an abstract of 'Measurement of the deflection of light by the sun's gravitational field during the eclipse of June 19th, 1936', *C.R. Moskau (N.S.)* **29,** 189–90. For more details of these and other measurements see von Kluber (1960).

Now eqn. (7.21) is the Schwarzschild line element which is the external solution to the field equations of general relativity. It is not surprising then that Cheng obtained equation (7.19). The question as to where he obtained the required transformation formulae was asked by Schild (1945) in his abstract of Cheng's note. There was in fact a small amount of literature on this topic, and it runs up to today.

The line element ds of equation (7.21) was first obtained from Einstein's field equations by Schwarzschild (1916) and a few months later by Droste (1917). Kottler (1918) derived this line element from certain postulates, including a requirement that the Newtonian equations were obtained as an approximation. This approach was used by Eisenhart (1920) to derive the Schwarzschild line element. The first elementary derivation seems to have been made by Phillips (1922). He used two fundamental hypotheses, the equivalence hypothesis ('Acceleration of the reference system is equivalent to gravitation') and the infinitesimal hypothesis ('In the infinitesimal neighbourhood of a world point (x,y,z,t) it is possible to choose a reference system with respect to which the restricted relativity is valid'). He also brought in the requirement of approximating to the Newtonian form at large distances. The same three postulates were used by Shimizu (1924).

It appears from this that Cheng was being elliptical when he asserted that the principle of equivalence sufficed to obtain the equivalent of the Schwarzschild solution, for local validity of special relativity was needed and also the Newtonian form as an approximation. I have found no other elementary deductions published before Cheng's note, but soon after Sommerfeld (1944) used an unpublished argument of W. Lenz to derive this line element. More recently Eriksson has given a similar derivation (and criticized Sommerfeld's argument) (Eriksson 1963–64, Eriksson and Yngstrom 1963) as has Tangherlini (1962). Such a derivation is to be found in some recent textbooks, but this type of argument has also been strongly criticised. This is the argument of Lenz as was given by Sommerfeld:

Consider a centrally symmetric gravitational field, e.g. that of the sun, of mass M, which may be regarded as at rest. Let a box K_∞ fall in a radial direction toward M. Since it falls freely, K_∞ is not aware of gravitation and therefore carries continually with itself the Euclidean metric valid at infinity. Let the coordinates measured within it be x_∞ (longitudinal, i.e. in the direction of motion), y_∞, z_∞ (transversal), and t_∞. K_∞ arrives at the distance r from the sun with the velocity v. v and r are to be measured in the system K of the sun, which is subject to gravitation. In it we use as coordinates r, θ, ϕ, and t. Between K_∞ and K there exist the relations of the special Lorentz transformation, where K plays the role of the system 'moving' with the velocity $v=\beta c$, K that of the system 'at rest'. The relations are

$$\begin{array}{ll}
\mathrm{d}x_\infty = \mathrm{d}r/\sqrt{(1-\beta^2)} & \text{(Lorentz contraction)} \\
\mathrm{d}t_\infty = \mathrm{d}t\cdot\sqrt{(1-\beta^2)} & \text{(Einstein dilatation)} \\
\left.\begin{array}{l} \mathrm{d}y_\infty = r\,\mathrm{d}\theta \\ \mathrm{d}\gamma_\infty = r\sin\theta\,\mathrm{d}\phi \end{array}\right\} & \begin{array}{l}\text{(invariance of the} \\ \text{transversal length).}\end{array}
\end{array}$$

Hence the Euclidean world line element

$$ds^2 = dx_\infty^2 + dy_\infty^2 + dz_\infty^2 - c^2\, dt_\infty^2 \tag{9}$$

passes over into

$$ds^2 = (1-\beta^2)^{-1}dr^2 + r^2(d\theta^2 + \sin^2\theta\, d\phi^2) - c^2 dt^2(1-\beta^2). \tag{9a}$$

The factor $1-\beta^2$, which occurs here twice, is meaningful so far only in connection with our specific box experiment. In order to determine its meaning in the system of the sun we write down the energy equation for K_∞, as interpreted by an observer on K. Let m be the mass of K_∞, m_0 its rest mass. The equation then is

$$(m-m_0)c^2 - GMm/r = 0 \tag{10}$$

At the left we have the sum of the kinetic energy in accord with Eq. (32.7) $[T = m_0c^2\{1/\sqrt{(1-\beta^2)}-1\} = (m-m_0)c^2]$ and of the (negative) potential energy of gravitation. The energy constant on the right was to be put equal to zero since at infinity $m=m_0$ and $r=\infty$. We have computed the potential energy from the Newtonian law, which we shall consider as a first approximation. We divide Eq. (10) by mc^2 and obtain then, since $m = m_0/\sqrt{(1-\beta^2)}$,

$$1-\sqrt{(1-\beta^2)} = \alpha/r \qquad \alpha = GM/c^2 = kM/8\pi \tag{10a}$$

... it follows from Eq. (10a) that $\sqrt{(1-\beta^2)} = 1-\alpha/r$

$$1-\beta^2 \approx 1-2\alpha/r$$

and hence, from Eq. (9a)

$$ds^2 = \frac{dr^2}{1-2\alpha/r} + r^2(d\theta^2 + \sin^2\theta d\phi^2) - c^2\left(1-\frac{2\alpha}{r}\right)dt^2. \tag{12}$$

(Sommerfeld 1952, pp. 313–14)

Sommerfeld's equation is the exact external solution to Einstein's field equations, the external Schwarzschild solution. The principle of equivalence enters in the third sentence when the effect of free fall is said to be the same as the effect of being in a gravitational field. Sommerfeld should have further specified this to be a homogeneous gravitational field since the lines normal to the equipotential lines converge towards the central body, in this case the Sun, instead of being parallel, and this convergence could be detected. Because of the equivalence the geometry which K observes to hold for K_∞ may be taken to be his own. The form of the geometry is specified by the Lorentz transformation, but needs the Newtonian potential to completely specify the content. Eriksson criticized Sommerfeld for making 'no clear distinction between coordinate differentials and measured length and time differentials'. (Eriksson and Yngstrom 1963, p. 327). Eriksson's derivation was designed to clarify the distinction between the dr etc. as measured intervals and differentials of the co-ordinates, but in doing so he opened himself to attack.

Sacks and Ball (1968)† have pointed out that although the derivations of

† Sacks and Ball also criticized a paper by Schiff (1960). They approve of Tangherlini's (1962) approach.

Lenz and Eriksson give a line element which looks like the Schwarzschild line element, their solution is in fact not an invariant and has no physical significance. In terms of Sommerfeld's presentation of Lenz's argument (it would be more rigorous to compare Sacks and Ball directly with p. 171 of Eriksson (1963–64), to determine the Lorentz contraction and the Einstein dilatation one requires a time-like and a space-like pair of events respectively. Substituting these into the Euclidean element produces a non-invariant expression because a line element is only invariant if it corresponds to the difference in co-ordinates of a single pair of events, and one cannot find a pair of events that is both time-like and space-like. More recently Rindler (1968) has given a counter-example to the Lenz type of argument by showing that it leads to absurd consequences for a particular example of a gravitational field. It also conflicts with early theories of gravity. The argument required that a gravitational field (strictly a homogeneous one) and a uniform acceleration were indistinguishable, which is the strong principle of equivalence (which entails the weak principle, i.e. the equaliy of inertial and gravitational masses). Therefore theories satisfying this and special relativity and that reduce to Newtonian theory, such as Nordström's second theory, should also give the Schwarzschild metric, whereas they manifestly do not.

What relevance has this to Einstein and the perihelion of Mercury? The answer is that, given his penchant for heuristic arguments, it was possible that if this type of argument had been valid Einstein could have used it to obtain the Schwarzschild metric before he had derived the field equations and thus found the perihelion advance. He did use the principle of equivalence to obtain an intermediate metric, and this will now be considered.

The argument was proposed by Einstein in 1911 (Einstein 1911, pp. 101–7). He compared two systems, K and K′. K is situated in a homogeneous gravitational field while K′ is gravitation free but accelerating along the z axis (and in the direction of increasing z) of a co-ordinate system where the z axis is the direction of the gravitational field lines of K (and the potential decreases with increasing z). Two points S_1 and S_2 are fixed a distance h apart on the z axis of K′. Viewed from a third system K_0, the systems K and K′ are considered to be at the same position on the z axis. Then the acceleration of K′ is said to be γ and the difference in the gravitational potential between S_1 and S_2 is γh. A representation of the situation is shown in Fig. 7.2. Radiation E_2 is emitted from S_2 towards S_1, and at this moment let the velocity of K′ relative to K_0 be zero. To a first approximation the radiation travelling the distance h from S_2 to S_1 at a velocity c takes a time h/c. After a time h/c, K_0 measures the velocity of S_1 to be v, where

$$v = u+at = 0+\gamma\cdot\frac{h}{c} = \frac{\gamma h}{c}. \tag{7.22}$$

The energy of this radiation is then measured to be different from E_2, and this value E_1 is related to E_2 by

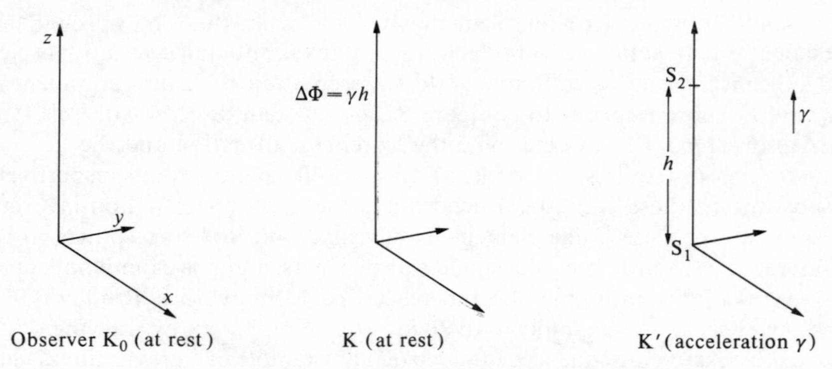

FIG. 7.2 Reference frames for the 1911 argument.

$$E_1 = E_2(1+v/c). \tag{7.23}$$

This is obtained from the transformation formula of special relativity where

$$\begin{aligned} E' &= E\left(\frac{1-v/c}{1+v/c}\right)^{\frac{1}{2}} \\ &= E\left(1-\frac{v}{c}\right)^{\frac{1}{2}}\left(1+\frac{v}{c}\right)^{-\frac{1}{2}} \\ &\approx E\left(1-\frac{v}{2c}\right)\left(1-\frac{v}{2c}\right) \\ &= E\left(1-\frac{v}{c}\right) \end{aligned}$$

to first order. E is measured in the stationary system and E' in the moving system. Thus $E = E_1$ and $E' = E_2$, and $E_1 = E_2(1-v/c)^{-1} = E_2(1+v/c)$ to first order. Substituting equation (7.22) into (7.23) gives

$$E_1 = E_2(1+\gamma h/c^2). \tag{7.24}$$

Now γh is the difference in gravitational potential between S_1 and S_2, and we can choose constants such that the potential Φ is zero at S_2, and thus $\Phi = \gamma h$. Then from equation (7.24)

$$E_1 = E_2\left(1+\frac{\Phi}{c^2}\right). \tag{7.25}$$

In these last steps, from equation (7.24) to (7.25) the validity of transferring from system K′ to system K is assumed by the principle of equivalence, which Einstein took in the form that all laws of physics were the same in K′ and in K (strong equivalence).

From equation (7.25) Einstein showed that the frequencies ν_1 and ν_2 of the radiation were related by

$$\nu_1 = \nu_2(1+\Phi/c^2). \qquad (7.26)$$

This yielded the result that the solar spectral lines when measured on earth were shifted in frequency by an amount Φ/c^2 or 2×10^{-6} towards the red. This was the first numerical prediction of the red shift, and Einstein noted that such an effect had been observed but ascribed to solar surface phenomena.

Einstein then noted that it seemed absurd to say that the frequency changed, where frequencies are measured as periods per second, since the number of periods emitted by S_2 must be the same as the number arriving at S_1. To remove this apparent absurdity required that the measured second had changed or that time intervals were a function of gravitational potential. Since frequency is the number of periods per unit time interval, we obtain from equation (7.26)

$$dt_1 = dt_2(1+\Phi/c^2)^{-1}$$

or

$$dt_1 = dt_2(1-\Phi/c^2). \qquad (7.27)$$

Thus in a gravitational field clocks go slow or time dilates. The velocity of light is also a function of potential and, from equation (7.26), if c_0 is the value when the potential is zero we have

$$c = c_0(1+\Phi/c_0^2). \qquad (7.28)$$

This is the relation that Einstein used in his 1912 gravitational theory, and is the same expression as that in equation (7.1). However, how was this relation exactly obtained? This is what Einstein wrote:

This [the relation (7.27)] has a consequence which is fundamental importance for our theory. For if we measure the velocity of light at different places in the accelerated, gravitation-free system K′, employing clocks U of idenical constitution, we obtain the same magnitude at all these places. The same holds good, by our fundamental assumption, for the system K as well. But from what has just been said we must use clocks of unlike constitution, for measuring time at a place which, relatively to the origin of the coordinates, has the gravitational potential Φ, we must employ a clock which—when removed to the origin of coordinates—goes $(1+\Phi/c^2)$ times more slowly than the clock used for measuring time at the origin of coordinates. If we call the velocity of light at the origin of coordinates c_0, then the velocity of light c at a place with the gravitation potential Φ will be given by the relation

$$c = c_0(1+\Phi/c^2).$$

The principle of the constancy of the velocity of light holds good according to this theory in a different form from that which usually underlies the ordinary theory of relativity. (Einstein 1911, pp. 106–7)

It is stated here both that the velocity of light is a function of Φ and that its value measured locally is constant. To measure the velocity of light c we send a beam of light along a path of length dl and measure the time of travel dt and calculate

$$c = \mathrm{d}l/\mathrm{d}t. \tag{7.29}$$

In a gravitational potential Φ the time interval dt is transformed according to equation (7.27). To maintain the constant value for measured c we could either transform dl in the same way as dt, and this we have no independent argument for, or use a clock that cancels out the effect of equation (7.27). We shall see below that Nordström took the first way in his second theory, specifically to keep the velocity of light constant. Einstein, however, took the second way. One has to use a clock that goes slowly by a factor $1+\Phi/c^2$ for then we have, from equations (7.27) and (7.29), to first order

$$c = \frac{\mathrm{d}l}{\mathrm{d}t(1+\Phi/c^2)(1-\Phi/c^2)} = \frac{\mathrm{d}l}{\mathrm{d}t}$$

and the measured value of c is independent of Φ. However, the velocity of light in a potential Φ_1, when measured by an observer in a potential Φ_2 who uses a clock that goes slow by a factor $(1+\Phi_2/c^2)$, is measured as

$$c = \frac{\mathrm{d}l}{\mathrm{d}t(1+\Phi_2/c^2)(1-\Phi_1/c^2).}$$

If the observer is in a zero potential $\Phi_2 = 0$ and $c = c_0$ when Φ_1 and Φ_2 are zero or equal, then this becomes

$$c = \frac{c_0}{(1-\Phi_1/c^2)} = c_0(1-\Phi_1/c^2)^{-1}$$

or

$$c = c_0(1+\Phi_1/c^2) \tag{7.30}$$

to first order which is Einstein's relation (and equation (7.28)). Thus the two paths offered above had the following difference; whereas both guaranteed constancy of the velocity of light as measured locally, only Nordström's path guaranteed it to be constant globally. In his theory special relativity was valid globally, there was no light deflection and no electromagnetic–gravitational interaction. In Einstein's theory special relativity was valid only locally, there was light deflection and interaction between electromagnetic and gravitational fields.

Einstein obtained the same results in 1911 as he had previously in 1907 when he had introduced the equivalence principle and derived an expression for the potential dependence of the velocity of light. In 1907 he discounted accelerative effects on measured times and lengths (Einstein 1907, p. 455) and was able to use the Lorentz transformations which were applicable to steady motions and were found in special relativity. He considered a system falling from rest with acceleration γ. In time t it attains a velocity $v=\gamma t$ and travels a distance x. The system drops through a gravitational potential difference $\Phi=\gamma x$. The Lorentz transformation of a time t to a time t' is

$$t' = (1-v^2/c^2)^{-\frac{1}{2}}(t-vx/c^2), \tag{7.31}$$

which, neglecting terms of order v^2/c^2, reduce to

$$t' = t-vx/c^2. \tag{7.31a}$$

Substituting $v=\gamma t$ gives

$$t' = t-\gamma xt/c^2. \tag{7.31b}$$

The gravitational potential is given by $\Phi=\gamma x$, and since the principle of equivalence states that the effects of free fall are the same as in a gravitational field we can substitute this into equation (7.31b) to give

$$t' = t(1-\Phi/c^2), \tag{7.31c}$$

which is the same as the 1911 result, as expressed in equation (7.27).

The Lorentz transformation for a distance x is

$$x' = (1-v^2/c^2)^{-\frac{1}{2}}(x-vt). \tag{7.32}$$

Neglecting terms of order v^2/c^2 gives

$$x' = x-vt \tag{7.32a}$$

This is merely the pre-relativistic Galilean transformation and there is no place for the gravitational potential. In this theory length does not contract in a gravitational field.

If the time dilation and length preservation of Einstein's early probing towards a gravitational theory are expressed in a space-time metric, one obtains an intermediate metric in comparison with the Schwarzschild metric of general relativity.

In the flat space-time Minkowski metric

$$\mathrm{d}s^2 = \mathrm{d}l^2 - c^2\,\mathrm{d}t^2 \tag{7.33}$$

we replace the velocity of light c by the expression obtained in equation (7.28). If c is now taken to denote the velocity of light in the absence of a gravitational field, the intermediate metric obtained by substituting equation (7.28) into equation (7.33) is

$$\mathrm{d}s^2 = \mathrm{d}l^2 - c^2(1+\Phi/c^2)^2\,\mathrm{d}t^2$$

or

$$\mathrm{d}s^2 = \mathrm{d}l^2 - c^2(1+2\Phi/c^2)\,\mathrm{d}t^2 \tag{7.34}$$

neglecting the term in c^{-4}. We can now insert the gravitational potential which for a body of mass M is given by $\Phi = -GM/r$ at a distance r from the body, where G is the universal constant of gravitation. The metric of equation (7.34) becomes

$$\mathrm{d}s^2 = \mathrm{d}l^2 - c^2(1 - 2GM/r^2 c^2)\,\mathrm{d}t^2. \tag{7.35}$$

The effect of this intermediate metric upon the orbit of a planet will now be derived. The method is due to Cheng and was mentioned earlier. The Lagrangian of a particle is given by

$$L = \tfrac{1}{2}(\dot{r}^2 + r^2\dot{\theta}^2) - V(r) \tag{7.36}$$

where the first term is the kinetic energy of a particle of unit mass moving in a plane orbit with co-ordinates (r, θ). The Newtonian potential $V(r)$ is given by

$$V(r) = -GM/r \tag{7.37}$$

where M is the mass of the central body. In the metric (7.35) the interval $\mathrm{d}r$ is unchanged but the interval $\mathrm{d}t$ is transformed to $\mathrm{d}t'$, where

$$\mathrm{d}t' = (1 - \lambda/r)\,\mathrm{d}t \text{ or } \mathrm{d}t' = \beta^{-1}\,\mathrm{d}t \tag{7.38}$$

where

$$\beta = 1 + \lambda/r \tag{7.39}$$

and

$$\lambda = GM/c^2 \tag{7.40}$$

From equation (7.39) we obtain

$$\partial\beta/\partial r = -\lambda/r^2 \tag{7.41}$$

Hence using equations (7.37) and (7.38) equation (7.36) is transformed to

$$\begin{aligned} L &= \tfrac{1}{2}(\beta^2\dot{r}^2 + r^2\beta^2\dot{\theta}^2) + GM/r \\ &= \tfrac{1}{2}\beta^2(\dot{r}^2 + r^2\dot{\theta}^2) + GM/r. \end{aligned} \tag{7.42}$$

Now

$$\frac{\partial L}{\partial r} = \beta\frac{\partial\beta}{\partial r}(\dot{r}^2 + r^2\dot{\theta}^2) + \tfrac{1}{2}\beta^2\cdot\dot{\theta}^2\cdot 2r - \frac{GM}{r^2} \tag{7.43}$$

and

$$\frac{\partial L}{\partial \dot{r}} = \tfrac{1}{2}\beta^2 2\dot{r} = \beta^2\dot{r}. \tag{7.44}$$

Then the equation of motion

$$\frac{\mathrm{d}}{\mathrm{d}t}\left(\frac{\partial L}{\partial \dot{r}}\right) - \frac{\partial L}{\partial r} = 0$$

becomes, from equations (7.43) and (7.44),

$$\frac{\mathrm{d}}{\mathrm{d}t}(\beta^2\dot{r})-\beta\dot{r}^2\frac{\partial\beta}{\partial r}-\beta\dot{\theta}^2r^2\frac{\partial\beta}{\partial r}-\beta^2\dot{\theta}^2r+\frac{GM}{r^2}=0. \tag{7.45}$$

From equation (7.42) we also obtain

$$\frac{\partial L}{\partial\theta}=0 \tag{7.46}$$

and

$$\frac{\partial L}{\partial\dot{\theta}}=\tfrac{1}{2}\beta^2\,2\dot{\theta}r^2=\beta^2r^2\dot{\theta}. \tag{7.47}$$

Then the equation of motion

$$\frac{\mathrm{d}}{\mathrm{d}t}\left(\frac{\partial L}{\partial\dot{\theta}}\right)-\frac{\partial L}{\partial\theta}=0$$

becomes, using equations (7.46) and (7.47)

$$\frac{\mathrm{d}}{\mathrm{d}t}(\beta^2r^2\dot{\theta})=0$$

or

$$\beta^2r^2\dot{\theta}=h=\text{constant}. \tag{7.48}$$

Putting $u=1/r$ we find from eqn. (7.48)

$$\begin{aligned}\dot{r}&=-\frac{1}{u^2}\frac{\mathrm{d}u}{\mathrm{d}t}=-\frac{1}{u^2}\frac{\mathrm{d}u}{\mathrm{d}\theta}\dot{\theta}\\&=-\frac{1}{u^2}\dot{\theta}\frac{\mathrm{d}u}{\mathrm{d}\theta}=-\frac{h}{\beta^2}\frac{\mathrm{d}u}{\mathrm{d}\theta}\end{aligned}$$

or, where $\dot{u}\equiv\mathrm{d}u/\mathrm{d}\theta$,

$$\dot{r}=-\frac{h}{\beta^2}\dot{u}. \tag{7.49}$$

Hence from eqn. (7.49)

$$\begin{aligned}\frac{\mathrm{d}}{\mathrm{d}t}(\beta^2\dot{r})&=\frac{\mathrm{d}}{\mathrm{d}t}\left\{\beta^2\left(\frac{-h}{\beta^2}\right)\dot{u}\right\}\\&=\frac{\mathrm{d}}{\mathrm{d}t}(-h\dot{u})\\&=-h\frac{\mathrm{d}}{\mathrm{d}t}\left(\frac{\mathrm{d}u}{\mathrm{d}\theta}\right)\\&=-h\frac{\mathrm{d}^2u}{\mathrm{d}\theta^2}\frac{\mathrm{d}\theta}{\mathrm{d}t}\\&=-h\ddot{u}\frac{h}{r^2\beta^2}\end{aligned}$$

where $\ddot{u} \equiv \mathrm{d}^2u/\mathrm{d}\theta^2$ and using equation (7.48)

$$= -\frac{h^2}{\beta^2} u^2\ddot{u}. \tag{7.50}$$

Using equations (7.20), (7.41) and (7.48) we obtain from equation (7.42)

$$-\frac{h^2}{\beta^2} u^2\ddot{u} - \beta \frac{h^2}{\beta^4} \dot{u}^2 \cdot -\lambda u^2 - \beta \cdot \frac{h^2u^4}{\beta^4} \frac{1}{u^2} \cdot -\lambda u^2 - \beta^2 \frac{h^2u^4}{\beta^4} \cdot \frac{1}{u} + GMu^2 = 0$$

$$-\left(\frac{h^2}{\beta^2}\right) u^2\ddot{u} + \left(\frac{h^2}{\beta^2}\right) u^2 \frac{\lambda}{\beta} \dot{u}^2 + \left(\frac{h^2}{\beta^2}\right) u^2 \cdot \frac{\lambda}{\beta} \mathrm{u}^2 - \left(\frac{h^2}{\beta^2}\right) u^2 \cdot u + GMu^2 = 0$$

$$\ddot{u} - \left(\frac{\lambda}{\beta}\right) \dot{u}^2 - \left(\frac{\lambda}{\beta}\right) u^2 + u - \frac{GM\beta^2}{h^2} = 0$$

or

$$\ddot{u} + u = \left(\frac{GM\beta^2}{h^2}\right) + \left(\frac{\lambda}{\beta}\right) u^2 + \left(\frac{\lambda}{\beta}\right) \dot{u}^2. \tag{7.51}$$

Now from equation (7.39) $\beta = 1 + \lambda u$. Therefore to first order in $1/c^2$, from

$$\beta^2 = 1 + 2\lambda u \qquad \text{and} \qquad \frac{\lambda}{\beta} = \lambda\beta^{-1} = \lambda$$

Therefore equation (7.51) becomes

$$\ddot{u} + u = \frac{GM}{h^2} + \left(\frac{2\lambda GM}{h^2}\right) u + \lambda u^2 + \lambda \dot{u}^2$$

or

$$\ddot{u} + u = k + 2k\lambda u + \lambda u^2 + \lambda \dot{u}^2 \tag{7.52}$$

where

$$k = GM/h^2 \tag{7.53}$$

The solution of $\ddot{u} + u = k$ is $u = k(1 + e \cos \gamma\theta)$. Substituting for u in equation (7.52) gives, to first order,

$$\begin{aligned} \ddot{u} + u &= k + 2k^2\lambda + 2k^2\lambda e \cos \gamma\theta + k^2\lambda + 2k^2\lambda e \cos \gamma\theta \\ &= k + 3k^2\lambda + 4k^2\lambda e \cos \gamma\theta. \end{aligned} \tag{7.54}$$

The particular integral of $\ddot{u} + u = A \cos \phi$ is

$$u = \tfrac{1}{2}A\phi \sin \phi.$$

Thus the last term in equation (7.54) gives an addition to u of

$$u_1 = 2k^2\lambda e \cdot \gamma\theta \cdot \sin \gamma\theta.$$

Hence the second approximation is

$$u = k(1+e\cos\gamma\theta+2k\lambda e\gamma\theta\sin\gamma\theta)+3k^2\lambda$$
$$= k\{1+e\cos(\gamma\theta-\delta\varpi)\}+3k^2\lambda \tag{7.55}$$

where

$$\delta\varpi = 2k\lambda\gamma\theta \ll 1.$$

Hence we obtain an ellipse with a perihelion that advances each revolution by an amount $\delta\varpi$, where

$$\frac{\delta\varpi}{\gamma\theta} = 2k\lambda$$
$$= 2\frac{GM}{h^2}\frac{GM}{c^2}$$
$$= 2\frac{GM}{GMa(1-e^2)}\frac{GM}{c^2}$$
$$= \frac{2GM}{ac^2(1-e^2)}: \tag{7.56}$$

general relativity gives for this quantity $3GM/ac^2(1-e^2)$, and so this theory gives two-thirds of the required result.

We can also calculate the light deflection due to this metric. This is actually treating the ray of light as composed of particles coming in from infinity and is thus similar to the Soldner analysis; the required condition is thus $h=\infty$, but the analysis is less fundamental than that of general relativity which obtains this condition by setting $ds=0$, the condition for the path of a light ray.

Setting $h=\infty$ gives from equation (7.53) $k=0$. Substituting in equation (7.52) gives

$$\ddot{u}+u = \lambda u^2+\lambda\dot{u}^2 \tag{7.57}$$

The solution of $\ddot{u}+u=0$ is $u=\cos\theta/R$. Substituting this into equation (7.57) gives

$$\ddot{u}+u = -\lambda\frac{\cos^2\theta}{R^2}-\lambda\frac{\sin^2\theta}{R^2}$$
$$= -\frac{\lambda}{R^2}(\cos^2\theta+\sin^2\theta) = -\frac{\lambda}{R^2}. \tag{7.58}$$

Therefore the solution of equation (7.57) is

$$u = \frac{\cos\theta}{R}-\frac{\lambda}{R^2}$$

or

$$\frac{1}{r} = \frac{\cos\theta}{R} - \frac{\lambda}{R^2}$$

or

$$R = r\cos\theta - \frac{\lambda}{R}r$$

or

$$R = x \pm \frac{\lambda}{R}(x^2+y^2)^{1/2}.$$

Thus

$$x = R \pm \frac{\lambda}{R}(x^2+y^2)^{1/2} \tag{7.59}$$

and this represents a slight deviation from the straight line $x = R$. Taking $y \gg x$ gives us the asymptotes, for which $x = R \pm (\lambda/R)y$. Therefore the total light deflection is

$$\frac{2\lambda}{R} = \frac{2GM}{c^2 R},$$

which is half the relativistic value and the same as Soldner would have found if he had not got a factor of 2 wrong. It is the value Einstein obtained in 1911 when he used Huyghen's principle, which was the more correct method to use as photons have no rest mass.

We have thus shown that this intermediate metric gives two-thirds the full perihelion advance and half the light deflection of the Schwarzschild metric. Schild (1960) has given perihelion advances and light deflections for a generalized metric, and his results agree with the above. The gravitational red shift is determined by the coefficient of dt in the metric, so our previously obtained result—that the 'full' shift is predicted—follows directly through. Whitrow and Morduch (1960, p. 792) have shown that these three results also occur in Einstein's gravitational theory of 1912.

The heuristic arguments of Phillips, Lenz, and others were designed to give both time dilatation and length contraction in order to give the Schwarzschild metric, which is (*cf.* equation (7.21))

$$\mathrm{d}s^2 = \left(1 - \frac{2\Phi}{c^2}\right)\mathrm{d}r^2 + r^2\mathrm{d}\theta^2 + r^2\sin^2\theta\,\mathrm{d}\phi^2 - c^2\left(1 + \frac{2\Phi}{c^2}\right)\mathrm{d}t^2. \tag{7.60}$$

This gives the full perihelion advance. Since the red shift is determined solely by the coefficient of dt, and this is as in the intermediate metric, this also gives the full red shift. The influence on the velocity of light may easily be seen by setting $\mathrm{d}s=0$ in equation (7.60). Choosing $\mathrm{d}\theta=\mathrm{d}\phi=0$, i.e. the radial solution, we obtain

$$0 = \left(1-\frac{2\Phi}{c^2}\right)\mathrm{d}r^2 - c^2\left(1+\frac{2\Phi}{c^2}\right)\mathrm{d}t^2$$

$$\mathrm{d}r^2\left(1-\frac{2\Phi}{c^2}\right) = \mathrm{d}t^2\cdot c^2\left(1+\frac{2\Phi}{c^2}\right)$$

or

$$\left(\frac{\mathrm{d}r}{\mathrm{d}t}\right)^2 = c^2\left(1+\frac{2\Phi}{c^2}\right)\left(1-\frac{2\Phi}{c^2}\right)^{-1}$$

$$\left(\frac{\mathrm{d}r}{\mathrm{d}t}\right) = c\left(1+\frac{\Phi}{c^2}\right)\left(1+\frac{\Phi}{c^2}\right)$$

or

$$\left(\frac{\mathrm{d}r}{\mathrm{d}t}\right)\bigg|_{\mathrm{d}s\,=\,0} = c\left(1+\frac{2\Phi}{c^2}\right).$$

Thus the velocity of light in a gravitational potential Φ is, if c_0 is the velocity when $\Phi=0$, given by

$$c = c_0(1+2\Phi/c_0{}^2). \tag{7.61}$$

The potential now has twice the effect it had in the intermediate metric as expressed in equation (7.28). Einstein had found in 1911 that the angle of deflection per unit path of a light ray in a gravitational potential Φ was given by the expression $(-1/c)(\partial s/\partial n')$, or using equation (7.30), by $(-1/c^2)(\partial\Phi/\partial n')$. The Schwarzschild metric then gives for $(-1/c)(\partial c/\partial n')$ the value $(-1/c)(2/c)(\partial\Phi/\partial n')$ or $(-2/c^2)(\partial\Phi/\partial n')$. This is twice the 1911 value and is the full general relativistic value; the rest of the 1911 derivation proceeds as before.

Einstein's 1911 derivation of the potential dependence of time intervals was revived by Schild (1962–63), who used it to argue that any theory that yields a gravitational red shift must be a curved space–time theory. He then rejected Whitehead's theory because it appeared to give a red shift but was in fact a flat space–time theory, charging the theory with unsatisfactory *ad hoc*ness in its red-shift prediction. This was important, for it was not until 1971 that Whitehead's theory was shown to be empirically refuted (Will 1971).

The theory of Nordström does not suffer from this argument as its metric is curved—or more specifically it is conformally flat. It has been mentioned that Nordström's theories involved a mass dependence on the gravitational potential. In developing his second theory in 1913 he expressed the inertial mass m of the electron in terms of its gravitational mass M_g, charge e, radius a, and the velocity of light c as

$$m = \frac{e^2+M_g^2}{8\pi c^2 a}.$$

Since e and M_g are constant, for small changes of potential one had $\delta m \propto (1/\delta a)$ (Nordström 1913, pp. 543–6, 546–9). Then from the relation $m/\Phi' = \text{constant}$, where Φ' was the external gravitational field, he obtained

$$a\Phi' = \text{constant}. \tag{7.62}$$

This result was then used against Einstein's blackbody radiation argument that has already been considered. Nordström then wrote

> The dependence of the length dimensions of bodies on Φ' motivates the question whether or not the progress of time in physical processes is also influenced by the gravitational potential. For a simple case we can answer the question without further ado. On account of the constancy of the velocity of light it is of course clear that the time in which a light signal is propagated from one end of a rod to the other, increases in the same proportion as the rod is lengthened. This time is therefore inversely proportional to the gravitational potential. (Nordström 1913, p. 546)

> From the example dealt with last it follows that the wavelength of a spectral line depends on the gravitational potential. A numerical calculation shows that wavelengths on the surface of the Sun must be about two millionths greater than from terrestial sources of light. Some of the other new theories of gravitation also give the same—perhaps even observable—displacement. (Nordström 1913, p. 549)

Nordström's theory yields the same shift as the Einstein theories. Its metric is of the form

$$ds^2 = \Phi'^2(dx^2+dy^2+dz^2-c^2\,dt^2) \tag{7.63}$$

which is conformally flat.

The red shift in the solar spectral lines was contentious in that some observers found no shift and others found one but could give plausible non-gravitational explanations. However, it was the same on the intermediate and Schwarzschild metrics, as well as on that of Nordström. The deflection of starlight was not measured until 1919, and moreover there was no evidence for estimating whether any such deflection was likely. The classic paper of Dyson, Eddington, and Davidson (1920, p. 291) which gave the 1919 results listed as 'likely' alternatives, no deflection (though no mention of Nordström), the 0″.87 Newtonian value, and the 1″.75 of general relativity. This left the perihelion advance as the sole prediction that could have told Einstein that his 1911 line interval was not the full one required and that length contraction was also needed. Yet this was the one predicton that was left uncalculated and unmentioned, the only one that could have shown the inadequacy of the intermediate metric. It has been argued in this chapter that Seeliger's hypothesis had defused Mercury's perihelion advance of its anomalous status and that gravitational theorists were no longer interested in incorporating it into their theories. This is weakened slightly when one realizes that the other two effects were not previously anomalous either, and that the red shift, like the perihelion advance, could be explained either by non-Newtonian gravitational methods or 'Newtonian' solar surface effects. In fact Forbes has shown how, after the 1919 light deflection results, the existence of the Einstein red shift was assumed and non-gravitational occurrences were used to explain the remaining discrepancies. In 1912 the two-thirds perihelion advance could have been assumed and circumsolar light matter used to explain the remaining discrepant one-third. There were controversies concerning these non-gravitational red shift explanations so that there was room to assume that another type of explanation was appropriate, whereas there was no controversy about Seeliger's hypothesis until Freundlich published his paper in early 1915. However, Seeliger's hypothesis was completely *ad hoc* as regards the density of the disturbing matter and could therefore accommodate any advance. Possibly Einstein's remark that he thought Seeliger's hypothesis was an 'open door' should be taken seriously; perhaps he wished to keep his perihelion predictions until he had obtained the full value, especially since he knew that his theories up to the late 1915 general relativity were theoretically imperfect. On this account he published the red shift predictions and light deflection predictions because he could not tell whether they were incomplete or not.

That Einstein did not use metrics in discussion is not relevant, since all one needs is the transformation relations for time and distance measurements. One does find explicit statements of metrics in a few places. In 1912 Einstein wrote:

For the material point moving in the static field of gravitation without effect of external forces

$$\delta\{\int H\,\mathrm{d}t\} = 0 \text{ is valid, or}$$

$$\delta\{\int\sqrt{(c^2\mathrm{d}t^2-\mathrm{d}x^2-\mathrm{d}y^2-\mathrm{d}z^2)}\} = 0.$$

(Einstein 1912b, p. 458)

In the 1913 Vienna lecture we find

For the 'natural' four dimensional element ds we obtain in the approximate case considered here

$$\mathrm{d}s = \sqrt{(-\mathrm{d}x^2-\mathrm{d}y^2-\mathrm{d}z^2+g_{44}\mathrm{d}t^2)}$$

where

$$g_{44} = c^2(1-\frac{\aleph}{4\pi}\int\frac{\rho_0}{r}\,\mathrm{d}v).$$

(Einstein 1913, p. 1260)

From this he proceeded to mention the time dilatation and light deflection. In the long Berlin paper of 1914 he gave explicitly our intermediate metric:

In the case of the Newtonian approximation there emerges for the naturally measured distance neighbouring a point in space–time

$$ds^2 = \sum_{\mu,\nu} g_{\mu\nu}\, dx_\mu\, dx_\nu = -dx^2 - dy^2 - dz^2 + (1+2\phi)dt^2.$$

(Einstein 1914a, p. 1084)

From this he derived the red-shift and light deflection. The metric of the *Entwurf* theory as given by Droste (1915, p. 1000) is much more complicated than the intermediate metric.

However, the perihelion advance is a more complicated effect than the other two predictions. For a start it applies to real planets that are not point particles. If the new theory gave the full advance, and it was to be correct, it also had to give the same perturbation results as Newtonian celestial mechanics. In the discussion following the 1913 Vienna lecture Max Born probed this aspect:

Born: I must direct a question at Herr Einstein, namely how quickly is the gravitational action propagated in your theory? That it happens with the velocity of light does not seem evident to me, it must be a very complicated connection.

Einstein: It is extraordinarily easy to write down the equations for the case where the disturbances which are inserted into the field are infinitely small. Then the *g*s are distinguished from those which were present without that disturbance only by an infinitely small amount; the disturbances are propagated then with the same velocity as that of light.

Born: But it is indeed very complicated for large disturbances?

Einstein: Yes, then it is a mathematically complicated problem. In general it is difficult to find the exact solutions of the equations as they are not linear. (Einstein 1913, p. 1266)

von Laue (1917*a*)† gave as one of his reasons for not accepting general relativity straight away the belief that the perihelion calculation would not apply to extended bodies.

Even the question of whether the perihelion advance is a first– or second-order effect in general relativity needs a subtle discussion as Duff (1974) has recently shown.

Considerations of this type suggest that simple perihelion advance calculations could run into trouble when applied to the existing situation. Einstein may have foreseen this when he gave such calculations for the red shift and the light deflection but not for the perihelion. We know that the difficulties do not in fact arise and we can derive all three effects from the metric, but there was less justification for this when trying to establish the theory.

The principle of equivalence was a permanent keystone of Einstein's

† Cited by Guth (1970, p. 185). This review article of the Nordström theory is not reprinted or mentioned in von Laue's *Gesammelte Schrifte*. The relevant passage reads:

This agreement between two individual numbers [the perihelion prediction of Einstein and the Newcomb anomaly], achieved under conditions which cannot be arbitrarily altered, so that it seems uncertain whether the suppositions (specifically the assumption of two mass points) are fulfilled with sufficient accuracy, does not seem to be a sufficient reason, even though it is note-worthy, to change the whole physical conception of the world to the full extent as Einstein did in his theory.

gravitational theories. Others were less enthusiastic about it. Abraham, while accepting the results of the Eötvös experiment (equality of inertial and gravitational masses), refused to take the inductive step leading to the principle.† Silberstein (1918) put forward a theory in which the principle of general covariance was retained but the principle of equivalence was rejected. It was rejected because it placed 'gravitation on an entirely exceptional and privileged footing'. Silberstein also felt empirical misgivings about general relativity. The deflection of light rays had yet to be tested. The advance of the perihelion was 'most vitally conditioned' by g_{44}, the coefficient of the $c^2 dt^2$ term of the Schwarzschild metric, which also as we have seen responsible for the gravitational red shift. In 1914 Freundlich had tested the red shift in the spectral lines of certain double stars and had found support for Einstein's prediction, but in 1917 St John had found no red shift in the solar spectrum. Silberstein chose St John's results as being superior on the grounds that they were more numerous and that the Sun's mass was better known than the masses of Freundlich's stars. Now no red shift implied a constant g_{44} and in turn a reduced perihelion motion, and consequently the success of the latter could not be taken as confirming the theory. Moreover it was the g_{44} that assured Newtonian planetary motion as a first approximation. Silberstein repeated his views at the meeting of the Royal Astronomical Society in 1919 which discussed general relativity; Lindemann and Jeffreys argued against him.‡ In his theory Silberstein used a potential

$$\Phi = \frac{M}{R}\cot\left(\frac{r}{R}\right) \tag{7.64}$$

which was the same as Neumann had used in 1886 (see equation (7.15)). He took the g_{ij} as constant *in vacuo*, with g_{22} and g_{33} as functions of R, the curvature of space. He obtained a perihelion regression of 7″ per century for Mercury's perihelion, which Jeffreys thought could still be covered by Seeliger's hypothesis—a view which he retracted in 1919 (see Chapter 4).

7.8. The principle of covariance and the perihelion advance

The principle of covariance as given in Einstein's 1916 paper on the foundations of general relativity states that:

The general laws of nature are to be expressed by the equations which hold good for all

† Cited by Mehra (1974, p. 6) referring to Abraham (1912*a*). In this paper Abraham referred to the theory that Einstein was developing but which was still in a sketchy stage as a *fata morgana*.

‡ *Mon. Not. r. astr. Soc.* **80,** 111–7 (1919). On p. 108 Sir Oliver Lodge slightly anticipated the Lenz argument when discussing the coefficients of the Schwarzschild metric:

The second term, more clearly written as $2M/r/c^2$, is the gravitational potential at any point divided by the square of the velocity of light. The numerator is the squared velocity of free fall from infinity. And as a beam of light has really fallen from infinity, the expression at once assumes a common-sense aspect.

systems of coordinates, that is, are co-variant with respect to any substitutions whatever (generally co-variant). (Einstein 1916)

The principle was closely connected (but not the same as) the move to express physical laws in terms of tensors, and Einstein held it as fundamental to his theory. Nevertheless it was abandoned by him in the *Entwurf* period, and only when he returned to it was the final version of general relativity obtained. In contrast to this the principle of equivalence was never dropped.

For our present purposes it is sufficient to note that this principle has no empirical significance since any law can be put into covariant form. This was pointed out by Kretschmann (1917) and had been known to Ricci and Levi-Civita in 1901 (see Whittaker 1953, p. 159). The importance of this principle according to Einstein himself lay in its heuristic use, or more accurately, in its heuristic use coupled with a strong requirement of simplicity. Generally covariant formulations of Newtonian gravitational theory were given by Cartan (1923, 1924) and by Friedrichs (1927). The subject has recently been discussed by A. Trautmann et al. (1965) and Havas (1964).

I have in the preceding discussion pointed out the lack of concern in the papers of the gravitational theorists that were of relevance about the advancing perihelion of Mercury. I have pointed out the crucial role this could have played in distinguishing between Nordström's theories and Einstein's theories, the role of Seeliger's hypothesis in disguising the anomalous nature of the advance, and possible difficulties in giving a simple calculation of the advance like those Einstein gave for the red shift and the light deflection. However, it would be presumptious to conjecture why exactly Einstein did not publish perihelion calculations until the final theory was obtained, for his reasons are not discernible in his papers and this study has been of published sources only and not of archival material. I have limited myself to pointing out difficulties and possibilities.

7.9. The Brans–Dicke theory and Mercury's perihelion

The story of general relativity and the perihelion of Mercury does not end here, for the advance and its explanation have recently been the subject of controversy. This arose as a result of a complete change in the situation regarding gravitational theories which occurred in the early 1960s. Before then only the three classical tests of general relativity were capable of confirming that theory owing to experimental limitations. Since, for example, Whitehead's theory gave exactly the same predictions for these three effects Whitehead (1922), the confirmation was not strong. However, in the early 1960s increases in the sensitivity of experimental techniques gave the possibility of testing more subtle effects, and in the wake of this change there arose many new gravitational theories. The change may be observed by comparing two *Proceedings of the International School of Physics Enrico*

Fermi, one—*Evidence for Gravitational Theories*—published in 1962 and the other—*Experimental Gravitation*—published in 1974. Only in 1971 was Whitehead's theory refuted when Will (1971) showed that it predicted isotropic inertial masses but anisotropic gravitational masses. The effect would have been concentrated in the direction of the centre of the Galaxy with its high density of matter which would have resulted in a tidal effect in the measured value of the gravitational constant G, with a 12-h period, of 200 times the magnitude of the observed value.

Brans and Dicke (1961) put forward a scalar–tensor theory of gravity which tried to incorporate Mach's principle into general relativity. This principle—that inertial forces are gravitational effects of distant matter—is only imperfectly satisfied in general relativity, and Brans and Dicke tried to remedy this situation by adding a scalar field to the usual tensor field which would represent this interaction. This Brans–Dicke theory is thus a scalar–tensor theory and contains a coupling constant ω, which is a measure of the coupling of the two fields (and where 'in any sensible theory ω must be of the general order of magnitude of unity' (Brans and Dicke 1961)). The Brans–Dicke theory predicts a perihelion advance equal to

$$\delta\varpi = \left(\frac{4+3\omega}{6+3\omega}\right) \times \text{(value predicted by general relativity)} \qquad (7.65)$$

In order for the observed advance to agree with the relativistic prediction to an accuracy of 8% or less Dicke required $\omega \geqslant 6$. For $\omega = 6$ the prediction is 39.4″ which is 3.2″ less than the accepted figure of 42.6″ ± 0.9″. Dicke saw the possibility of ascribing the difference to an oblateness of the Sun, and consequently set up experiments to test for oblateness (Dicke and Goldenberg 1967; see also Dicke 1974, Roxburgh 1974). These experiments convinced Dicke that a sufficient oblateness existed to allow a coupling constant of value 6. There were difficulties with this, for this oblateness was too great to be caused by the observed rate of spin of the Sun, and this required a high spin rate of the solar core which seemed to imply instability. The high oblateness did find confirmation in solar models accounting for low values of observed neutrino flux investigated by Demarque, Menzel, and Sweizert (1973). However, Hill and Stebbins (1975) have shown that, by giving a more rigorous definition of the solar edge, they can account for the Dicke–Goldenberg results by attributing them to excessive equatorial brightness. Hill and Stebbins found an oblateness of $18.4 \pm 12.5 \times 10^{-3}$ arcsec compared with the Dicke–Goldenberg value of $86.6 \pm 6.6 \times 10^{-3}$ arcsec.

Meanwhile other tests were confirming general relativity. Shapiro *et al.* (1971) tested the round-trip time delay of an electromagnetic wave by bouncing radar signals off Mercury and Venus, and found a constant λ equal to 1.01 ± 0.02 where Brans–Dicke predicted 0.93 and general relativity 1.00. Anderson *et al.* (1972) also confirmed the relativistic values using the Mariner

6 and Mariner 7 missions. Counselman *et al.* (1974) measured the deflection of radio waves in the solar gravitational field and confirmed general relativity. A similar experiment using microwaves by Fomalont and Sramek (1975) gave a value to the deflection which requires the Brans–Dicke coupling constant to satisfy the inequality $\omega > 23$. This brings the perihelion advance to a value close to the usual general relativistic one and leaves little room for solar oblateness. This has been confirmed, and the inequality made more restrictive, by a recent experiment on lunar laser ranging performed by Williams *et al.* (1976). The experiment was designed to test a consequence of the scalar–tensor theory known as the Nordtvedt effect in which the gravitational self-energy of a body contributes more to the gravitational mass than to the inertial mass (Nordtvedt 1968). The effect is not observable at the laboratory level (e.g. Dicke's experiments on the equivalence principle which updated those of Eötvös) but is for large bodies. The resulting mass correction is greater for the Earth than for the Moon, so that there would be an excessive acceleration of the Earth towards the Sun which would manifest itself as a variation in the Earth–Moon distance. This was tested by lunar laser ranging using cubical mirrors left on the Moon by American astronauts and to a 70 per cent confidence level the results require the coupling constant to satisfy $\omega > 29$. Thus the Brans–Dicke theory is hardly distinguishable from general relativity, in particular for its prediction of the advance in the perihelion of Mercury.

REFERENCES

ABBOT, C.G. (1898). The Newtonian Lucretius. *Annu. Rep. Smithson. Inst.*, pp. 139–60.

ABRAHAM, M. (1912*a*). Nochmals Relativität und Gravitation, Bemerkungen zu A. Einsteins Erwiderung. *Annln Phys. (Lpz.)* **39,** 444–8.

—— (1912*b*). Zur Theorie der Gravitation. *Phys. Z.* **13,** 1–4.

—— (1912*c*). Das elementargesetz der Gravitation. *Phys. Z.* **13,** 4–5.

—— (1912*d*). Das Gravitationsfeld. *Phys. Z.* **13,** 793–7.

—— (1914*a*). Neuere Gravitationstheorie. *Jahrb. Radioakt. Elektron.* **11,** 470–520.

—— (1914*b*). Die neue Mechanik. *Scientia* **15,** 8–27 (French translation: pp. 10–29 of the *Supplement* (see p. 24)).

ADAMS, J.C. (1847). On the perturbations of Uranus. *Mem. r. astr. Soc.* **16,** 427–460.

—— (1853). On the secular variation of the moon's mean motion. *Mon. Not. r. astr. Soc.* **14,** 59–62.

—— (1859) On the secular variation of the eccentricity and inclination of the moon's orbit. *Mon. Not. r. astr. Soc.* **19,** 206–8.

AIRY, G.B. (1833) Report on astronomy. *Rep. br. Ass.* 125–89.

—— (1849) Corrections of the elements of the moon's orbit. *Mem. r. astr. Soc.* **17,** 21–58.

—— (1855). Observations of daily temperatures referred to the meridian of the Sun's body. *Astr. Nachr.* **39,** cols. 337–40.

—— (1884). *Gravitation* (2nd edn), Section IV. Macmillan, London.

ANDERSON, J.D., ESPOSITO, P.B., MARTIN, W. and MUHLEMAN, D.O. (1972). Measurement of General Relativistic time delay with Mariners 6 and 7. *Space Res. XII* **2,** 1623–30.

ANDING, E. (1905) Über Koordinaten und Zeit. *Encykl. math. Wiss.* **6**(2), 3–15.

ANON. (1859–60). A supposed new interior planet. *Mon. Not. r. astr. Soc.* **20,** 98–101.

D'ARREST, H.L. (1854). Ueber die ungleiche Vertheilung der Wärme auf der Sonnenoberfläche. *Astr. Nachr.* **37,** cols. 263–8.

VON ASTEN, E. (1878). Untersuchungen über die Theorie der Enkeschen Cometen. II. Resultate aus den Erscheinungen 1819–1875. *Mem. Acad. imp. Sci. St Petersbourg* **26,** 1–125.

AUWERS, A. (1891). Der Sonnendurchmesser und der Venusdurchmesser nach den Beobachtungen an den Heliometern der deutschen Venus-Expeditionen. *Astr. Nachr.* **128,** cols. 361–76.

BABINET, M. (1846). Mémoire sur les nuages ignés du soleil considérés comme des masses planétaires. *C.R. Acad. Sci. Paris* **22,** 281–6.

—— (1848). Sur la position actuelle de la planète située au delà de Neptune, et provisoirement nommée Hypérion. *C.R. Acad. Sci. Paris* **27,** 202–8.

BACKLUND, O. (1908–1911). *La Comète d'Encke 1891–1908.* Académie impériale des sciences, St Petersbourg. (See in particular Sur l'acceleration du mouvement moyen, *Fascicule III* (1911), p. 40f.).

BARNARD, F.A.P. (1857). On the theory which attributes the zodiacal light to a nebulous ring surrounding the earth. *Am. J. Sci.* **21,** 217–37, 399–401.

Barton, E.H. (1916). *Nature (Lond.)* **97,** 461–2.
—— (1917). *Nature (Lond.)* **99,** 45.
Bauschinger, J. (1884). Zur Frage über die Bewegung der Mercurperihels. *Astr. Nachr.* **109,** cols. 27–32.
Behacker, M. (1913). Der freie Fall und die Planetenbewegung in Nordströms Gravitationstheorie. *Phys. Z.* **14,** 989–92.
Bertrand, G. (1922). La loi de Riemann, le périhélie de Mercure et la déviation de la lumière. *C.R. Acad. Sci. Paris* **174,** 1687–9.
Bertrand, J. (1873) Theorème relatif au mouvement d'un point attiré vers un centre fixe. *C.R. Acad. Sci. Paris* **77,** 849–53.
—— (1890). *Leçons sur la théorie mathematique de l'electricité.* Gauthier-Villars, Paris. Paris.
Birkenhead, Earl of (1961) *The Prof in two worlds*, p. 92. Collins, London.
Bohr, N. (1913). On the constitution of atoms and molecules. *Phil. Mag.* **26,** 1–25.
Born, M. and Wolf, E. (1959). *Principles of optics.* Pergamon Press, London.
Bottlinger, K.F. (1912). Die Erklärung der empirische Glieder der Mondbewegung durch die Annahme einer Extinktion der Gravitation im Erdinnern. *Astr. Nachr.* **191,** cols 147–50.
—— (1914). Zur Frage der Absorption der Gravitation. *Sitzungsber. Math.-Naturwiss. Kl. Bayer. Akad. Wiss. Münch.* 223–9.
Brans, C. and Dicke, R.H. (1961). Mach's Principle and a relativistic theory of gravitation. *Phys. Rev.* **124,** 925–35.
Brasch, F.E. (1929). Einstein's appreciation of Simon Newcomb. *Science* **69,** 248–9.
Brewster, D. (1845). *N. Br. Rev.* **4,** 227.
Brown, E.W. (1897). On the mean motions of the lunar perigee and node. *Mon. Not. r. astr. Soc.* **57,** 332–41, 566.
—— (1903). On the verification of the Newtonian law. *Mon. Not. r. astr. Soc.* **62,** 396–7.
—— (1904). On the degree of accuracy of the new lunar theory. *Mon. Not. r. astr. Soc.* **64,** 530.
—— (1910). On the effects of certain magnetic and gravitational forces on the motion of the Moon. *Am. J. Sci.* **29,** 529–39.
Bucherer, A.H. (1922). Gravitation und Quantentheorie. *Annln. Phys.* **68,** 1–10, 545–50.
Burton, C.V. (1909). A modified theory of gravitation. *Phil. Mag. (Ser. 6)* **17,** 71–113.
Buys-Ballot, C.H.D. (1846). Über den Einfluss der Rotation der Sonne auf die Temperatur unserer Atmosphäre. *Annln. Phys. Chem.* **68,** 205–213.
—— (1847). *Changements de température dépendents du soleil et de la lune mis en rapport avec le pronostic du temps déduits d'observations Neerlandaises de 1729 à 1846*, Kemink, Utrecht.
—— (1860). Iets over een Ring om de Zon, door de Astronomie vermoed en door de Meteorologie nader aangewezen. *Versl. Meded. k. Akad. Wet. Afd. Natuurk.* **10,** 110 (see pp. 113–118).
Caldonazzo, B. (1913). Traiettorie dei Raggi Luminosi e dei punti materiali nel campo gravitazionale. *Nuovo Cim., Ser. 6* **5,** 267–300.
Cartan, E. (1923). *Annls Éc. Norm.* **40,** 325.
—— (1924). *Annls Éc. Norm.* **41,** 1.
Cartwright, D.E. (1977). Oceanic tides. *Rep. Prog. Phys.* **40,** 665–708.
Cassini, J.D. (1730). Découverte de la lumière céleste qui paroit dans le zodiaque. *Mem. Acad. r. Sci. 1666–1699* **8,** 121–209.
Challis, J. (1859). The force of gravity. *Phil. Mag. (Ser. 4)* **18,** 442–51.

—— (1861). On the planet within the orbit of Mercury, discovered by M. Lescarbault. *Proc. Camb. phil. Soc.* **1**(15), 219–22.

—— (1863). A theory of the zodiacal light. *Phil. Mag. (Ser. 4)* **25,** 117–25, 183–9.

CHAMBERS, G.F. (1889). *Handbook of descriptive and practical astronomy (4th edn.),* Vol. 1, Book 1, Chap. 3. Clarendon Press, Oxford.

CHANDLER, P. (1975). Clairaut's critique of Newtonian attraction: some insights into his philosophy of science. *Ann. Sci.* **32,** 369–78.

CHILDREY, J. (1661). *Brittania Baconica,* p. 183.

CLAIRAUT, A.C. (1745). Du système du monde dans les principes de la gravitation universelle. *Hist. Mem. Acad. r. Sci.* **58,** 329–64 (read November 15, 1747).

CLARK, R.W. (1971). *Einstein: the life and times.* World Publishing, New York.

CLAUSIUS, R. (1877). Ableitung eines neuen elektrodynamisches Grundgesetzes. *J. reine angew. Math.* **82,** 85–130.

CLERKE, A.M. (1887). *A popular history of astronomy during the nineteenth century.* (2nd edn). Adam and Charles Black, Edinburgh.

CLIFFORD, W.K. (1876). On the space theory of matter. *Proc. Camb. phil. Soc.* **2,** 157.

COMSTOCK, D.F. (1910). A neglected type of relativity. Astronomical consequences of the assumption that the velocity of light depends on the source velocity. *Phys. Rev.* **30,** 267.

COOKE, C. (1882). Bjerknes's hydrodynamical experiments. *Engineering* **33,** 23–5, 147–8, 191–2.

Cosmos (1852). **21,** 471.

Cosmos (1859). **15,** 609.

COSTER, H.G.C. and SHEPANSKI, J.R. (1969). Gravito-inertial fields and relativity. *J. Phys. A.* **2,** 22–7.

COUNSELMANN, C.C. *et al.* (1974). Solar gravitational deflection of radio waves measured by very-long-baseline interferometry. *Phys. Rev. Lett.* **33,** 1621–3.

DARWIN, G. (1905). The analogy between Lesage's theory of gravitation and the repulsion of light. *Proc. R. Soc. A* **76,** 387–410.

—— (1908). The tides. *Encycl. math. Wiss.* **6**(1), 3–83 (transl. in G. Darwin, *Scientific papers,* vol. 4, p. 185; Chap. E, Tidal friction and speculative astronomy; S41, History).

—— (1916). Introduction to dynamical astronomy. *Sci. Pap. (Cambridge)* **5,** 9–15.

DAS, A. (1957). On the perihelion shift in conformally flat space–time. *Nuovo Cim. (Ser. 10)* **6,** 1489–90.

DEHALU, M. (1926). Le mouvement du périhélie de Mercure deduit de certaines lois de gravitation. *Bull. Cl. Sci. Acad. r. Belg.* **12,** 381–93, 639.

DELAUNAY, C. (1870). Sur la constitution physique de la lune. *C.R. Acad. Sci. Paris* **70,** 57–61.

DEMARQUE, P., MENZEL, J.G. and SWEIZERT, A.V. (1973). Rotating solar models with low neutrino flux. *Astrophys. J.* **183,** 997.

DE SITTER, W. (1911). On the bearing of the principle of relativity on gravitational astronomy. *Mon Not. r. Astr. Soc.* **71,** 388–415, errata pp. 524, 603, 716.

—— (1913*a*). On absorption of gravitation and the moon's longitude. *Proc. k. ned Akad. Wet., Sect. Sci.* **15,** 808–39.

—— (1913*b*). The secular variations of the elements of the four inner planets. *Observatory* **36,** 296–303.

—— (1913*c*). Ein astronomisches Beweis für die Konstanz der Lichtgeschwindigkeit. *Phys. Z.* **14,** 429.

—— (1913*d*). Uber die Genauigkeit, innerhalb welcher die Unabhängigkeit der

Lichtgeschwindigkeit von der Bewegung der Quelle behauptet werden kann. *Phys. Z.* **14,** 1267.

—— (1914). Remarks on Mr. Woltjer's paper concerning Seeliger's hypothesis. *Proc. k. ned. Akad. Wet., Sect. Sci.* **17,** 33–7.

—— (1916*a*). On Einstein's theory of gravitation and its astronomical consequences. *Mon. Not. r. astr. Soc.* **76,** 699–728.

—— (1916*b*). On Einstein's theory of gravitation and its astronomical consequences. *Mon. Not. r. astr. Soc.* **77,** 155–84.

—— (1917). On Einstein's theory of gravitation and its astronomical consequences. *Mon. Not. r. astr. Soc.* **78,** 3–28.

DE TUNZELMANN, G.W. (1910). *A treatise on electrical theory and the problem of the universe*, Chap. 18, p. 362. Charles Griffin, London.

DICKE, R.H. (1974). Relativity and the solar oblateness. *Proc. Int. School of Physics Enrico Fermi, Course 56, Experimental Gravitation*, pp. 200–34. Academic Press, New York.

DICKE, R.H. and GOLDENBERG, H.M. (1967). Solar oblateness and General Relativity. *Phys. Rev. Lett.* **18,** 313–16.

DORLING, J. (1974). Henry Cavendish's deduction of the electrostatic inverse square law from the result of a single experiment. *Stud. Hist. Phil. Sci.* **4,** 327–48.

DREYER, J.L.E. (1953). *A history of astronomy from Thales to Kepler*, p. 421. Dover Publications, New York.

DROSTE, J. (1915). On the field of a single centre in Einstein's theory of gravitation. *Proc. k. ned. Akad. Wet., Sect. Sci.* **17,** 998–1011.

—— (1917). The field of a single centre in Einstein's theory of gravitation and the motion of a particle in that field. *Proc. k. ned. Akad. Wet., Sect. Sci.* **19,** 197–215.

DUFF, M.J. (1974). On the significance of perihelion shift calculations. *Gen. Relativity Gravitation* **5**(4), 441–52.

DUNCOMBE, R.L. (1958). Motion of Venus 1750–1949. *Astr. Pap. am. Ephem.* **16,** 1–258.

DYSON, F.W., EDDINGTON, A.S., and DAVIDSON, C. (1920). A determination of the deflection of light by the sun's gravitational field, from observations made at the total eclipse of May 29, 1919. *Phil. Trans. R. Soc. London, Ser. A* **220,** 291–333.

EINSTEIN, A. (1907). Über das Relativitätstheorie und die aus demselbengezogenen Folgerungen. *Jahrb. Radioakt. Elektron.* **4,** 411–62.

—— (1911). Über den Einfluss der Schwerkraft auf die Ausbreitung des Lichtes. *Annln Phys. (Lpz.)* **35,** 898–908. (Translation: H.A. Lorentz *et al.* (1952). *The principle of relativity*, pp. 97–108. Dover Publications, New York.)

—— (1912*a*). Lichtgeswindigkeit und Statik des Gravitationsfeldes. *Annln Phys. (Lpz.)* **38,** 355–69.

—— (1912*b*). Zur theorie des Statischen Gravitationsfeldes. *Annln Phys. (Lpz.)* **38,** 443–58.

—— (1913). Zum gegenwartigen Stande des Gravitationsproblems. *Phys. Z.* **14,** 1249–66.

—— (1914*a*). Die formale Grundlage der allgemeinen Relativitätstheorie. *Sitzungsber. k. preuss. Akad. Wiss.* Pt. 2, 1030–85.

—— (1914*b*). Zum Relativitäts-Problem. *Scientia* **15,** 337–48 (French transl., *Suppl.* 139–50).

—— (1914*c*). Zur Theorie der Gravitation. *Vierteljahrsschr. Naturforsch. Ges. Zurich* **59**(2), 4–6.

—— (1915). Zur allgemeinen Relativitätstheorie; Zur allgemeinen Relativitätstheorie (Nachtrag); Erklärung der Perihelbewegung des Merkur aus der allgemeinen

Relativitätstheorie; Die Feldgleichungen der Gravitation. *Sitzungsber. k. preuss. Akad. Wiss.* Pt. 2, 778–86, 799–801, 831–9, 844–7.
—— (1916). Die Grundlage der allgemeinen Relativitätstheorie. *Annln Phys. (Lpz.)* **49,** 769–822 (trans. in Lorentz *et al.* 1953, p. 117).
—— (1918). Bemerkung zu E. Gehrckes Notiz "Über den Äther". *Verhandl. dtsch. phys. Ges.* **20,** 261.
EINSTEIN, A. and FOKKER, A.D. (1914). Die nordstromsche Gravitationstheorie vom Standpunkt des absoluten Differentialkalkuls. *Annln Phys. (Lpz.)* **44,** 321–8.
—— and GROSSMANN, M. (1913). Entwurf einer verallgemeinerten Relativitätstheorie und einer Theorie der Gravitation. *Z. Math. Phys.* **62,** 225–61.
EISENHART, L.P. (1920). The permanent gravitational field in the Einstein theory. *Proc. Natl. Acad. Sci. (Wash.)* **6,** 678–82.
ERIKSON, K.E. (1963–4). A simple derivation of Schwarzschild's metric. *Ark. Fys.* **25,** 167–73.
ERIKSON, K.E. and YNGSTROM, S. (1963). A simple approach to the Schwarzschild solution. *Phys. Lett.* **5**(2), 119, 327.
FARR, C.C. (1897). On an objection to LeSage's theory of gravitation. *Trans. Proc. N.Z. Inst.* **30,** 118–20.
FITZGERALD, G.F. (1894). Boltzmann on Maxwell. *Nature (Lond.)* **49,** 381–2.
FOMALONT, E.B. and SRAMEK, R.A. (1975). A confirmation of Einstein's General Theory of Relativity by measuring the bending of microwave radiation in the gravitational field of the sun. *Astrophys. J.* **199,** 749–55.
FONTENROSE, R. (1973). The discovery of Vulcan. *J. Hist. Astr.* **4,** 145–58.
FÖPPL, A. (1897). Uber eine mögliche Erweiterung des Newton'schen Gravitations Gesetzes. *Münch. Sitzungsber.* **27,** 93–9.
FORBES, E.G. (1961). A history of the solar red shift problem. *Ann. Sci.* **17,** 129–64.
FOWLER, A., *et al.* (1919). Discussions on the theory of relativity. *Mon. Not. r. Astr. Soc.* **80,** 96–118.
FOX, J.G. (1962). Experimental evidence for the second postulate of Special Relativity. *Am. J. Phys.* **30,** 297–300.
—— (1965). Evidence against emission theories. *Am. J. Phys.* **33,** 1–17.
FREUNDLICH, E. (1913). Zur Frage der Konstanz der Lichtgeschwindigkeit. *Phys. Z.* **14,** 835–8.
—— (1915) Über die Erklärung der Anomalien im Planeten-System durch die Gravitalionswirkung interplanetarer Massen. *Astr. Nachr.* **201,** cols. 49–56.
FRIEDRICHS, K. (1927). Eine invariante Formulierung des Newtonschen Gravitationsgesetzes und des Grenzüberganges vom Einsteinschen zum Newtonschen Gesetz. *Math. Ann.* **98,** 566–575.
GANS, R. (1905). Gravitation und Elektromagnetismus. *Phys. Z.* **6,** 803–5.
—— (1912). Ist die Gravitation elektromagnetischen Ursprungs? In *Festschrift Heinrich Weber*, pp. 75–94. Teubner, Leipzig.
GAUSS, C.F. (1867). *Werke*, Vol. 5. Königliche Gesellschaft der Wissenschaften zu Göttingen, Göttingen.
GAUTIÈR, A. (1817). *Essai historique sur le problème des trois corps, Part 1, Chap. 2, Theorie de la lune de Clairaut. Mouvement de l'apogée de la lune.* Courcier, Paris.
GEHRCKE, E. (1916). Zur Kritik und Gesichte der neueren Gravitations-theorien. *Annln Phys. (Lpz.)* **51,** 119–24.
—— (1918). Über den Äther. *Verhandl. dtsch. phys. Gesell.* **20,** 165–9.
—— (1919). Zur Diskussion über den Äther. *Verhandl. dtsch phys. Ges.* **21,** 67–8.
GERBER, P. (1898). Die raümlich und zeitliche Ausbreitung der Gravitation. *Z. Math. Phys.* **43,** 93–104.

—— (1902). Die Fortpflanzungsgeschwindigkeit der Gravitation. *Programmabhandlung des städtische Realgymnasiums zu Stargard i Pommerania.*

—— (1917). Die Fortpflanzungsgeschwindigkeit der Gravitation. *Annln Phys. (Lpz.), Ser. 4* **52,** 415–41.

VON GLEICH, G. (1923). Die allgemeine Relativitätstheorie und das Merkurperihel. *Annln Phys. (Lpz.)* **72,** 221–35.

GOLDSTEIN, H. (1950). *Classical mechanics*, p. 19. Addison-Wesley, Reading, Mass.

GOULD, B.A. (1850). On Kirkwood's analogy. *Am. J. Sci.* **10,** 26–31.

GRANT, R. (1852). *History of Physical Astronomy*. Henry G. Bohn, London.

GRAVÉ, D. and SOKOLOFF, G. (1926). Sur le mouvement du perihelie de Mercure. *Mém. Cl. Sci. phys. math. Acad. Sci. Ukraine* **5,** 1–11.

GROSSER, M. (1962). *The discovery of Neptune*. Harvard University Press, Cambridge, Mass.

GROSSMANN, E. (1921). Die Bewegung des Merkurperihels nach Arbeiten Newcombs. *Astr. Nachr.* **214,** cols. 41–54.

GÜRSEY, F. (1953). Gravitation and cosmic expansion in formal space–time. *Proc. Camb. phil. Soc.* **49,** 285–91.

GUTH, E. (1970). Contribution to the history of Einstein's geometry as a branch of physics. In *Relativity* (eds. M. Carmeli, S.I. Fickler, and L. Witten), pp. 161–208. Plenum, New York.

GUTHNIK, P. (1913). Astronomische Kriterien für die Unabhängigkeit der Fortpflanzungsgeschwindigkeit des Lichtes von der Bewegung der Lichtquelle. *Astr. Nachr.* **195,** cols 265–270.

GUTHRIE, F. (1870). On approach caused by vibration. *Phil. Mag.* **39,** 309, and **40,** 345–354.

HALL, A. (1894). A suggestion in the theory of Mercury. *Astr. J.* **14,** 49–51.

HANSEN, P.A. (1856). Sur la figure de la lune. *Mon. Not. r. astr. Soc.* 24, 29–89.

—— (1857). *Tables de la lune construite d'après le principe newtonien de la gravitation universelle*, British Government, London.

—— (1864). Darlegung der theoretische Berechnung der in den Mondtafeln angewandten Storungen. *Abh. Math.-Phys. kl. K.-Sächs. Ges. Wiss. (Lpz.)* **6,** 91–498; **7,** 1–399.

—— (1871). Ueber die Bestimmung der Figur des Mondes, in Bezug auf Aufsätze der Herren Newcomb und Delaunay darüber. *Sitzungsber. sächs. Akad. Wiss. Lpz. Math.-Naturwiss. Kl.* **23,** 1–12.

HANSON, N.R. (1962). Leverrier: the zenith and nadir of Newtonian mechanics. *Isis* **53,** 359–78.

HARKNESS, W. (1891). The solar parallax and its related constants including the figure and density of the earth. *Washington Observations for 1885*, Appendix 3, p. 102.

HARLEY, T. (1885). *Moon lore*, pp. 227–57. Sonnenschein, London.

HARRISON, E.R. (1974). Why the sky is dark at night. *Phys. Today* **27,** 30–6.

HARVEY, A.L. (1964). The principle of equivalence. *Ann. Phys. (N.Y.)* **29,** 383–90.

—— (1965*a*). Brief review of Lorentz-covariant scalar theories of gravitation. *Am. J. Phys.* **33,** 449–60.

—— (1965*b*). *Ann. Phys. (N.Y.)* **31,** 240–1.

HATTENDORF, K. (ed.) (1876). *Schwere, Elektricität und Magnetismus*, (Lectures by Riemann), p. 327. Rümpler, Hannover.

HAVAS, P. (1964). Four-dimensional formulations of Newtonian mechanics and their relation to the Special and General Theory of Relativity. *Rev. mod. Phys.* **36,** 938–65.

HELMHOLTZ, H. (1847). On the conservation of force; reprinted in J. Tyndall and W. Francis (eds.) (1853). *Sci. Mem.* **7,** 114–62.

—— (1870). Ueber die Bewegungsgleichungen der Elektricität für ruhende leitende Körper. *Crelle's J.* **72,** 57–129.

—— (1872). Ueber die theorie der Elektrodynamik (transl.). *Phil. Mag.* **44,** 530–7.

VON HEPPERGER, J. (1888). Über die Fortpflanzungsgeschwindigkeit der Gravitation. *Sitzungsber. k. Akad. Wiss. math. naturwiss. Cl., Wien,* **97,** 337–62.

HERMANN, A. (ed.) (1968). *Albert Einstein/Arnold Sommerfeld Briefwechsel.* Schwabe and Co., Basel/Stuttgart.

HERSCHEL, J. (1851). *Outlines of astronomy* (4th edn), p. 538. Longman, Brown, Green, and Longmans, London.

HICKS, W.M. (1880). On the problem of two pulsating spheres in a fluid. I. *Proc. Camb. phil. Soc.* **3,** 276–85.

—— (1880–83). On the problem of two pulsating spheres in a fluid. II. *Proc. Camb. phil. Soc.* **4,** 29–35.

—— (1895). Presidential Address to the Mathematics and Physical Science Section. *Rep. Br. Assoc.* 595–606.

HILL, G.W. (1891). Determination of inequalities of the moon's motion which are produced by the figure of the earth. *Astr. Pap. am. Ephem. naut. Alm.* **3,** 201–344.

HILL, H.A. and STEBBINS, R.T. (1975). The intrinsic visual oblateness of the sun. *Astrophys. J.* **200,** 471–83.

HOFFMANN, B. (1972). Einstein and tensors. *Tensor (New Ser.)* **26,** 157–62.

HOLZMÜLLER, G. (1870). Uber die Anwendung der Jacobi-Hamilton'schen Princips und das Weber'sche Gesetz. *Z. Math. Phys.* **15,** 69–91.

HOUGH, S.S. and HALM, J. (1910). On the systematic motions of the Bradley stars. *Mon. Not. r. astr. Soc.* **70,** 568–88.

HOVGAARD, W. (1931–2). Ritz's electrodynamic theory. *J. Math. Phys.* **11,** 218–54.

HOWSE, D. (1980). *Greenwich time*, Oxford University Press, Oxford.

HUNTINGTON, E. (1923). *Earth and sun: an hypothesis of weather and sunspots.* Yale University Press, New Haven, Conn.

INGERSOLL, A.P. and SPIEGEL, E.A. (1971). Temperature variation and the solar oblateness. *Astrophys. J.* **163,** 375–81.

JAKI, S. (1969). *The paradox of Olbers' paradox*, Chap. 9. Herder and Herder, New York.

JAMMER, M. (1954). *Concepts of space* (1st edn.), p. 167 ff. Harvard University Press, Cambridge, Mass.

JEANS, J.H. (1904). A suggested explanation of radioactivity. *Nature (Lond.)* **70,** 101.

—— (1919). *Problems of cosmogony and stellar dynamics*, p. 286. Cambridge Univ. Press, Cambridge.

JEFFREYS, H. (1916). The secular perturbations of the four inner planets. *Mon. Not. r. astr. Soc.* **77,** 112–8.

—— (1918). The secular perturbations of the inner planets. *Phil. Mag.* **36,** 203–5

—— (1919). On the crucial test of Einstein's theory of gravitation. *Mon. Not. r. astr. Soc.* **80,** 138–54.

—— (1921). *Science* **54,** 248.

JENKINS, B.G. (1878). Vulcan and Bode's law. *Nature (Lond.)* **19,** 74.

JONES, G. (1855). Cause of the zodiacal light. *Am. J. Sci. (Ser. 2)* **20,** 138–9.

—— (1856). Historical introduction. *Observations on the zodiacal light*, United States Japan Expedition, Washington, D.C.

KAI-CHIA CHENG (1945). A simple calculation of the perihelion of Mercury from the principle of equivalence. *Nature (Lond.)* **155,** 574.

KELVIN, LORD (1904). *Baltimore Lectures*, p. 274. Clay, London.
KILLING, W. (1885). Die Mechanik in den Nicht-Euklidischen Raumformen. *J. reine angew. Math.* **98,** 1–48.
KIRKWOOD, D. (1850). On a new analogy in the periods of the primary planets. *Am. J. Sci. (Ser. 2)* **9,** 395–9.
—— (1852). On certain analogies in the solar system. *Am. J. Sci.* **14,** 210–19.
—— (1864). On certain harmonies of the solar system. *Am. J. Sci.* **38,** 1–18.
VON KLUBER, H. (1960). The determination of Einstein's light-deflection in the gravitational field of the sun. *Vistas Astr.* **3,** 45–77.
KOTTLER, F. (1918). Über die physikalischen Grundlagen der Einsteinschen Gravitationstheorie. *Annln Phys. (Lpz.)* **56,** 401–62.
KOURGANOFF, M.V. (1941). La part de la mécanique céleste dans la découverte de Pluton. *Bull. Astr.* **12,** 147–330.
KRETSCHMANN, E. (1917). Über den physikalischen Sinn der Relativitäts-Postulat, A. Einsteins neue und seine ürsprungliche Relativitätstheorie. *Annln Phys. (Lpz.)* **53,** 575–614.
LANCZOS, C. (1974). *The Einstein decade*, p. 210. Elek Science, London.
LANGE, F.A. (1881). *History of materialism*, Vol. 3, p. 225ff. (transl. E.C. Thomas). English and Foreign Philosophical Library, London.
LAPLACE, P.S. (1805). *Mécanique céleste*, Vol. 4, Livre 10, Chap. 7, Sect. 22. Courcier, Paris.
LARMOR, J. (ed.) (1902). *The Scientific Writings of the late George Francis FitzGerald*, pp. 311–16. Longmans, Green, London.
—— (1916). *Nature (Lond.)* **97,** 321, 421.
—— (1917). *Nature (Lond.)* **99,** 44–5.
VON LAUE, M. (1917). Die Fortpflanzungsgeschwindigkeit der Gravitation. Bemerkungen zur gleichnamigen Abhandlungen von P. Gerber. *Annln Phys. (Lpz.)* **53,** 214–16.
—— (1917*a*). Die Nordströmsche Gravitations-theorie. *Jahrb. Radioakt. Elektron.* **14,** 263–313.
—— (1920). Historisch-Kritisches über die Perihelbewegung des Merkur. *Naturwissenschaften* **8,** 735–6.
LEAHY, A.H. (1889). On the pulsations of spheres in an elastic medium. *Trans. Camb. phil. Soc.* **14,** 45–62.
LEBEDEV, P. (1902). Die physikalische Ürsachen der Abweichungen vom Newton'schen Gravitationsgesetze (trans.). *Astrophys. J.* **16,** 155.
LEHMANN-FILHÉS, R. (1885). Ueber die Bewegung einer Planeten unter der Annahme einer sich nicht momentan fortpflanzenden Schwerkraft. *Astr. Nachr.* **110,** cols. 209–20.
—— (1895). Ueber die Säcularstörung der Länge des Mondes unter der Annahme einer sich nicht momentan forpflanzenden Schwerkraft. *Sitzungsber. math.-phys. Kl. Bayer Akad. Wiss. München* **25,** 371–422.
LENARD, P. (1936–7). *Deutsche Physik*. Lehmanns, Berlin.
LENSE, J. (1917). Das Newtonsche Gesetz in nichteuklidischen Raumen. *Sitzungsber. k. Akad. Wiss. Wien (Math.-Naturwiss. Kl.)* **126,** 1037–63.
LE VERRIER, U.J.J. (1849). Nouvelles recherches sur les mouvements des planètes. *C.R. Acad. Sci. Paris* **29,** 1–3.
—— (1859*a*). Theorie du mouvement de Mercure. *Ann. Observ. imp. Paris (Mém.)* **5,** 1–196 (also published as *Recherches astronomîques*, Vol. 5).
—— (1859*b*). Lettre de M. Le Verrier à M. Faye sur la théorie de Mercure et sur le mouvement du périhélie de cette planète. *C.R. Acad. Sci. Paris* **59,** 379–83.

—— (1876). Examen des observations q'on à presenteés, à diverse époques, comme pouvent appartenir aux passages d'une planète intra-mercurielle devant le disque du soleil. *C.R. Acad. Sci. Paris* **83,** 583–9, 621–4, 647–50, 719–23.

LÉVY, M. (1890*a*). Sur l'application des lois électrodynamiques au mouvement des planètes. *C.R. Acad. Sci. Paris* **110,** 545–51.

—— (1890*b*). Sur les diverse théories de l'électricité. *C.R. Acad. Sci. Paris* **110,** 740–2.

LIAIS, E. (1860). Sur la nouvelle planète annoncée par M. Lescarbault. *Astr. Nachr.* **52,** cols. 369–78.

—— (1881). *L'espace céleste.* (2nd edn) Garnier Frères, Paris.

LIEBMANN, H. (1902). Die Kegelschnitte und die Planetenbewegung im nichteuklidischen Raum. *Ber. Verhandl. k. sächs. Ges. Wiss. Lpz. (Math.-Phys. Kl.)* **54,** 393–423.

LINDEMANN, F.A. (1888). Über molekular-Physik. Versuch einer einheitlichen dynamischen Behandlung der physikalische und chemischen Kraften. *Schr. phys.-ökon. Ges. Königsberg* **29,** 31–81.

—— (1915). *Nature (Lond.)* **95,** 203–4, 372.

LINDEMANN, F.A. and BURTON, C.V. (1917). *Nature (Lond.)* **98,** 349.

LITTLEWOOD, D.E. (1953). Conformal transformations and kinematical relativity. *Proc. Camb. phil. Soc.* **49,** 90–6.

LODGE, O. (1917). *Nature (Lond.)* **99,** 104.

LORBERG, H. (1878). Ueber das Grundgesetz der Elektrodynamik. *Annln Phys. Chem.* **8,** 599–607.

LORENTZ, H.A. (1900). Considerations on gravitation. *Proc. Sect. Sci. (Amst.)* **2,** 559–74.

—— (1910). Alte und neue Frage der Physik. *Phys. Z.* **11,** 1234–57.

LOWELL, P. (1915). Memoir on a trans-Neptunian planet. *Mem. Lowell Obs.* **1,** 1–105.

LYTTLETON, R.A. (1953). *The comets and their origin.* Cambridge University Press, Cambridge.

—— (1963). Introduction to N.B. Richter, *The nature of comets* (transl. A. Beer). Methuen, London.

MACH, E. (1904). *Die Mechanik in ihrer Entwicklung Historisch-Kritisch Dargestellt* (5th edn), Brockhaus, Leipzig.

—— (1911). *History and root of the principle of the conservation of energy* (transl. P.E. Jourdain). Open Court, Chicago. (From *Die Gesichte und die Wurzel der Satzes von der Erhaltung der Arbeit*, Prague, 1872.)

—— (1960). *The science of mechanics* (6th edn), p. 368. Open Court, La Salle.

MAIN, R. (1853). In *Encyclopaedia Britannica* (8th edn), Vol. 4, p. 31.

MAJORANA, Q. (1920). On gravitation, theoretical and experimental researches. *Phil. Mag.* **39,** 488–504.

MARCHAND, E. (1895). Observations de la lumière zodiacale, faites á l'observatoire du Pic-Midi. *C.R. Acad. Sci. Paris* **121,** 1134–6.

MAXWELL, J.C. (1865). A dynamical theory of the electromagnetic field. *Phil. Trans.* **155,** 459–512 (see § 82, pp. 492–3).

—— (1873). *Treatise on electricity and magnetism*, Vol. 2. Clarendon Press, Oxford.

—— (1875). Atom. In *Encyclopaedia Britannica* (9th edn).

—— (1890). On Faraday's lines of force. In *Scientific papers*, Vol. 1. Cambridge University Press, Cambridge.

MEADOWS, A.J. (1970). *Early solar physics*, p. 10. Clarendon Press, Oxford.

MEHRA, J. (1974). *Einstein, Hilbert, and the theory of gravitation.* D. Reidel, Dordrecht.

MIE, G. (1913). Grundlagen einer Theorie der Materie, Chap. 5, Die Gravitation. *Annln Phys. (Lpz.)* **40,** 25–63.

MINKOWSKI, H. (1908). Die Grundgleichungen für die elektromagnetischen Vorgänge in bewegten Körpern. *Nachr. k. Ges. Wiss. Göttingen (Math.-Phys. Kl.)* 53–111.

MONK, W. (ed.) (1972). *The journals of Caroline Fox*, p. 175. Elek, London.

MOROZOV, N. (1908). *Zh. russ. khim. Ova.* **40,** 23–35.

MUNK, W.H., and MACDONALD, G.J.F. (1960). *The rotation of the earth.* Cambridge University Press, Cambridge.

NEUMANN, C. (1874). Über die den Kraften Elektrodynamischen Ursprungs zu zuschreibenden Elementargesetze. *Abh. Math.-Phys. Kl. k. sächs. Ges. Wiss. (Lpz.)* **10,** 419–524.

—— (1886). Ausdehnung der Kepler'schen Gesetze auf den Fall, dass die Bewegung auf einer Kugelfläche Stattfindet. *Ber. Verhandl. k. sächs. Ges. Wiss. Lpz. (Math.-Phys. Kl.)* **38,** 1–2.

—— (1896). *Allgemeine Untersuchungen über das Newton'sche Princip der Fernwirkungen mit besonder Rucksicht auf die Elektrischen Wirkungen*, p. 1. Teubner, Leipzig.

NEWCOMB, S. (1860). On the supposed intra-Mercurial planets. *Astr. J.* **6,** 162–3.

—— (1868). On Hansen's theory of the physical constitution of the moon. *Am. J. Sci. (Ser. 2)* **46,** 376–8.

—— (1882). Discussion and results of observations on transits of Mercury from 1677 to 1881. *Astr. Pap. am. Ephem. naut. Alm.*, **1,** 367–487. U.S. Govt. Printing Office, Washington, D.C.

—— (1895). The elements of the four inner planets and the fundamental constants of astronomy. *Suppl. am. Ephem. naut. Alm. 1897.* U.S. Govt. Printing Office, Washington, D.C.

—— (1895–8). *Astr. Pap. am. Ephem. naut. Alm.* **6**. U.S. Govt. Printing Office, Washington, D.C.

—— (1902). Astronomy. *Encyclopaedia Britannica* (10th edn).

—— (1903). *Reminiscences of an astronomer*. Harper, London.

—— (1908). La théorie du mouvement de la lune. *Rev. gén. sci. pures appl.* **19,** 686–91.

—— (1912). Researches in the motion of the moon. *Astr. Pap. am. Ephem. naut. Alm.* **9,** 226–7. U.S. Govt. Printing Office, Washington, D.C.

NEWTON, I. (1934). *Principia* (transl. A. Motte; revised by F. Cajori). University of California Press, Berkeley.

NORDSTRÖM, G. (1912). Relativitätsprinzip und Gravitation. *Phys. Z.* **13,** 1126–9.

—— (1913). Zur Theorie der Gravitation vom Standpunkt der Relativitätsprinzip. *Annln Phys. (Lpz.)* **42,** 533–54.

—— (1914). Die Fallgesetze und Planeten bewegung in der Relativitätstheorie. *Annln Phys. (Lpz.)* **43,** 1101–10.

NORDTVEDT, K., JR. (1968). Equivalence principle for massive bodies. *Phys. Rev.* **169,** 1014–6, 1017–25.

NORTH, J.D. (1965). *The measure of the universe*, p. 20. Clarendon Press, Oxford.

NUMBERS, R.L. (1973). The American Kepler: Daniel Kirkwood and his Analogy. *J. Hist. Astr.* **4,** 13–21.

OPPENHEIM, S. (1895). Zur Frage nach der Fortpflanzungsgeschwindigkeit der Gravitation. *Jahresber. k. k. akad. Gymnas. Wien Schulj. 1894/95*, 3–28.

—— (1917). Zur Frage nach der Fortpflanzungsgeschwindigkeit der Gravitation. *Annln Phys. (Lpz.)* **53,** 163–8.

OPPOLZER, T. (1879*a*). Elemente des Vulkan. *Astr. Nachr.* **94,** cols. 97–9.

—— (1879*b*). *Astr. Nachr.* **94,** cols. 303–4.

O'RAHILLY, A. (1965). *Electromagnetic theory*. Dover Publications, New York.

PAVANINI, G. (1912). Prime consequenze d'una recente teoria della gravitazione. *Atti r. Accad. Lincei, Ser. 5* **21** (2), 648–55.

—— (1913). Prime consequenze d'una recente teoria della gravitazione: le disuguaglianze secolari. *Atti. r. Accad. Lincei, Ser. 5* **22** (1), 369–76.

PEARSON, K. (1885). On the motion of spherical and ellipsoidal bodies in fluid media. *Q.J. pure appl. Math.* **20,** 60–79, 184–211.

—— (1888–89). On a certain atomic hypothesis. Part II. *Proc. London math. Soc.* **20,** 38–63.

—— (1889). On a certain atomic hypothesis. *Trans. Camb. phil. Soc.* **14,** 71–120.

—— (1891). Ether squirts. *Am. J. Math.* **13,** 309–62.

PETERS, C.H.F. (1879). Some critical remarks on so-called intra-Mercurial planet observations. *Astr. Nachr.* **94,** cols. 321–40.

PHILLIPS, H.B. (1922). Note on Einstein's theory of gravitation. *J. Math. Phys.* **1,** 177–90.

PIAZZI SMYTH, C. (1857). On the zodiacal light. *Mon. Not. r. astr. Soc.* **17,** 204–5.

PIRANI, F.E. (1955). On the perihelion motion according to Littlewood's equations. *Proc. Camb. phil. Soc.* **51,** 535–7.

POINCARÉ, H. (1906). Sur le dynamique de l'electron. *Rend. Circ. mat. Palermo* **21,** 494–550.

—— (1908). La dynamique de l'électron. *Rev. gén. Sci. pures appl.* **19,** 386–402.

—— (1953). Les limites de la loi de Newton. *Bull. astron.* **17,** 121–269.

—— (1954). *Oeuvres*, Vol. 9. Gauthier-Villars, Paris.

POOR, C.L. (1905). The figure of the sun. *Astrophys. J.* **22,** 103–14.

—— (1908). *The solar system.* Murray, London.

—— (1921). The motions of the planets and the relativity theory. *Science* **54,** 30–4.

—— (1922). *Relativity versus gravitation.* Putnam, New York.

—— (1925). Relativity and the motion of Mercury. *Ann. N.Y. Acad. Sci.* **29,** 285–319.

—— (1930). Relativity and the law of gravitation. *Astr. Nachr.* **238,** cols. 165–70.

PRESTON, S.T. (1877). On some dynamical conditions applicable to LeSage's theory of gravitation. *Phil. Mag. (Ser. 5)* **4,** 206–13, 364–75.

—— (1895). Comparative review of some dynamical theories of gravitation. *Phil. Mag. (Ser. 5)* **39,** 145–59.

PRINGSHEIM, E. (1910). *Vorlesungen über der Physik der Sonne*, p. 334. Teubner, Leipzig.

PROCTOR, R.A. (1870). Note on the zodiacal light. *Mon. Not. r. astr. Soc.* **31,** 2–10.

—— (1872). The zodiacal light. In *Essays on astronomy*, pp. 163–75. Longmans, Green, London.

—— (1873). *The borderland of science.* Longmans, Green, London.

—— (1891). *Other worlds than ours* (4th edn), pp. 173–4. Longmans, Green, London.

PROKHOVNIK, S.J. (1970). Cosmological theory of gravitation. *Nature (Lond.)* **225,** 359–61.

PYENSON, L. (1976). Einstein's early scientific collaboration. *Hist. Stud. phys. Sci.* **7,** 83–124.

REYNOLDS, O. (1903). *The submechanics of the universe.* Cambridge University Press.

RIEMANN, B. (1876). *Gesammelte mathematische Werke* (ed. H. Weber). Teubner, Leipzig.

RINDLER, W. (1968). Counterexample to the Lenz–Schiff argument. *Am. J. Phys.* **36,** 540–4.

RITZ, W. (1908). Recherches critiques sur l'électrodynamique générale. *Ann. Chim. Phys. Ser. 8* **13,** 145–275. Reprinted in *Gesammelte Werke*, pp. 317–422, Gauthier-Villars, Paris, 1911.

—— (1909). Die Gravitation. *Scientia* **5,** 241–55. French translation: La gravitation.

Suppl. Fr. Scientia **5,** 152–65. Reprinted in *Gesammelte Werke*, pp. 462–77, 478–92, Gauthier-Villars, Paris, 1911.

ROXBURGH, I.W. (1974). Solar oblateness and the solar quadrupole moment. *Proc. Int. School of Physics Enrico Fermi, Course 56, Experimental Gravitation*, pp. 525–8. Academic Press, New York.

RUSSELL, H.N., DUGAN, R.S., and STEWART, J.Q. (1928). *Astronomy*. Ginn, Boston.

RYSÁNEK, A. (1888). Versuch einer dynamischen Erklärung der Gravitation. *Repert. Phys.* **24,** 90–114.

SACKS, W.M. and BALL, J.A. (1968). Simple derivations of the Schwarzschild metric. *Am. J. Phys.* **36,** 240–5.

SCHEIBNER, W. (1897). Ueber die formale Bedeutung des Hamiltonschen Princips und das Weber'sche Gesetz. *Ber. Verhandl. k. sächs. Ges. Wiss. Lpz. math.-phys. Kl.* **49,** 578–607.

SCHIFF, L.I. (1960). On experimental tests of the General Theory of Relativity. *Am. J. Phys.* **28,** 340–3.

SCHILD, A. (1945). *Math. Rev.* **6,** 241.

—— (1960). Equivalence principle and red-shift measurements. *Am. J. Phys.* **28,** 778–80.

—— (1962). Gravitational theories of the Whitehead type and the principle of equivalence. In *Evidence for gravitational theories: Proc. Int. School Phys. Enrico Fermi*, Course 20 (ed. C. Møller), pp. 69–115. Academic Press, New York.

—— (1962–63). The principle of equivalence. *Monist* **47,** 20–39.

SCHOTT, G.A. (1906). On the electron theory of matter and the explanation of fine spectrum lines and of gravitation. *Phil. Mag. (Ser. 6)* **12,** 21–9.

SCHUSTER, A. (1898*a*). A holiday dream. *Nature (Lond.)* **58,** 367.

—— (1898*b*). *Nature (Lond.)* **58,** 618.

SCHWARZSCHILD, K. (1916). Ueber das Gravitationsfeld einer Massenpunktes nach der Einsteinschen Theorie. *Sitzunsgsber. dtsch. Akad. Wiss. Berlin* 189–96.

SEEGERS, C. (1864). De motu perturbationibusque planetarum secundum logem electrodynamicam Weberianam solem ambientum. *Inaugural Diss.* Göttingen.

SEELIG, C. (1956). *Albert Einstein.* Staples Press London.

SEELIGER, H. (1895). Ueber das Newton'sche Gravitationsgesetz. *Astr. Nachr.* **137,** cols. 129–36.

—— (1896). Ueber das Newton'sche Gravitationsgesetz. *Sitzungsber. Math. Naturwiss. Kl. Bayer. Akad. Wiss. Münch.* **26,** 373–400.

—— (1906). Das Zodiakallicht und die empirischen Glieder in der Bewegung der innern Planeten. *Sitzungsber. Math. Naturwiss. Kl. Bayer. Akad. Wiss. Munch.* **36,** 595–622.

—— (1909). Sur d'application des lors de la nature à l'univers. *Scientia* **6** (suppl.), 89–107.

—— (1915). Über die Anomalien in der Bewegung der innern Planeten. *Astr. Nachr.* **201,** cols. 273–80.

—— (1917*a*). Bemerkung zu P. Gerbers Aufsatz "Die Fortpflanzungsgeschwindigkeit der Gravitation". *Annln Phys. (Lpz.)* **53,** 31–2.

—— (1917*b*). Weiters Bemerkungen zur "Fortpflanzungsgeschwindigkeit der Gravitation". *Annln. Phys. (Lpz.)* **54,** 38–40.

—— (1918). Bemerkung zu dem Aufsatze der Herrn Gehrcke "Über den Äther". *Verhandl. dtsch. phys. Ges.* **20,** 262.

SEXL, R.U. (1967). Theories of gravitation. *Fortschr. Phys.* **15,** 269–307.

SHAPIRO, I.I. *et al.* (1971). Fourth test of General Relativity: new radar result. *Phys. Rev. Lett.* **26,** 1132–35.

SHAPLEY, H. (1960). *Source book in astronomy 1900–1950*, p. 342. Harvard University Press, Cambridge, Mass.

SHAW, P.E. (1916*a*). The Newtonian constant of gravitation as affected by temperature. *Phil. Trans. r. Soc. Lond., Ser. A* **216,** 349–92.

—— (1916*b*). *Nature (Lond.)* **97,** 400–1.

—— (1917). *Nature (Lond.)* **98,** 350; **99,** 84–5, 165.

SHAW, P.E. and DAVY, N. (1923). The effect of temperature on gravitative attraction. *Phys. Rev., Ser. 2* **21,** 680–91.

SHIMIZU, T. (1924). An elementary deduction of Einstein's law of gravitation. *Jpn. J. Phys. (Trans.)* **3,** 187–95.

SILBERSTEIN, L. (1918). General Relativity without the equivalence hypothesis. *Phil. Mag.* **36,** 94–128.

—— (1924). *The theory of relativity* (2nd edn), p. 298. Macmillan, London.

SMART, W.M. (1965). *Spherical astronomy* (5th edn), p. 213. Cambridge University Press.

SOMERVILLE, M. (1836). *The connexion of the physical sciences* (3rd edn). Murray, London.

SOMMERFELD, A. (1952). *Lectures on theoretical physics. Vol. 3. Electrodynamics* (trans. E.G. Ramberg), pp. 313–15. Academic Press, New York.

SPENCER-JONES, H. (1954). Dimensions and rotation. In *The earth as a planet* (ed. G. Kuiper), p. 32. Chicago University Press, Chicago, Ill.

SPEZIALI, P. (ed.) (1972). *Albert Einstein, Michele Besso, Correspondence 1903–1955.* Hermann, Paris.

STALLO, J.B. (1960). *The concepts and theories of modern physics*, p. 87. Harvard University Press, Cambridge, Mass, (originally published in 1881 by Appleton, New York).

STEWART, B. (1881). On the possibility of the existence of intra-Mercurial planets. *Nature (Lond.)* **24,** 463–4.

STROH, A. (1882). On attraction and repulsion due to sonorous vibrations and a comparison of the phenomena with those of magnetism. *J. Soc. telegr. Eng. Electr.* **11,** 192–228, 293–300.

STRUVE, O. and HUANG, S.-S. (1957). Close binary stars. *Occas. Note r. astr. Soc.* **3** (19), 161–88.

SU-CHING KIANG (1946). Deflexion of light in the gravitational field without using Einstein geometry. *Nature (Lon.)* **157,** 842.

SURDIN, M. (1962). A note on time-varying gravitational potentials. *Proc. Camb. phil. Soc.* **58,** 550–3.

SWINGS, P. (1927). Les orbites quasi elliptiques, les potentials riemanniens et les forces centrales. *Bull. Cl. Sci. Acad. r. Belg.* **13,** 88–99.

TAIT, P.G. and STEELE, W.J. (1856). *A treatise on the dynamics of a particle*, p. 88. Macmillan, London.

TANGHERLINI, F.R. (1962). Postulational approach to Schwarzschild's exterior solution with application to a class of interior solutions. *Nuovo Cim.* **25,** 1081–1105.

TAYLOR, W.B. (1876). Kinetic theories of gravitation. *Smithson. Rep. (Wash.)*, pp. 205–82.

THOMAS, J.S.G. (1917). *Nature (Lond.)* **99,** 405.

THOMSON, W. (1873). On the ultramundane corpuscles of LeSage. *Phil. Mag. (Ser. 4)* **45,** 321–32.

TISSERAND, F. (1872). Sur le mouvement des planètes autour du soleil, d'après la loi electro dynamique de Weber. *C.R. Acad. Sci. Paris* **75,** 760–3.

—— (1882). Notice sur les planètes intra-mercurielles. *Annu. Bur. Longit.*, 729–72.

—— (1890). Sur les mouvements des planètes, en supposant l'attraction représentée par l'une des lois électrodynamiques de Gauss ou de Weber. *C.R. Acad. Sci. Paris* **110,** 313–15.

—— (1896). *Mécanique céleste*, Vol. 4, p. 509. Gauthier-Villars, Paris.

Todd, G.W. (1917). *Nature (Lond.)* **99,** 5–6, 104–5.

Tolman, R.C. (1917). *The theory of the relativity of motion.* University of California Press, Berkeley.

Trautmann, A., Pirani, F.A.E. and Bondi, H. (1965). *Lectures on General Relativity.* Prentice Hall, Englewood Cliffs, N.J.

Turner, H.H. (1900). On the brightness of the corona of January 22, 1898. *Mon. Not. r. astr. Soc.* **61** (Appendix), 4–12.

Vierteljahrsschr. astr. Ges. (1902). **37,** 187. Darwin's lecture is reprinted on p. 202.

Wacker, F. (1906). Über Gravitation und Elektromagnetismus. *Phys. Z.* **7,** 300–2.

—— (1909). Über Gravitation und Elektromagnetismus. *Inaugural Diss.* Tübingen.

Waff, C.B. (1975). Alexis Clairaut and his proposed modification of Newton's inverse-square law of gravitation. In *Avant, avec, aprés Copernic.* 31e semaine de Synthèse, 1–7 Juin, 1973, pp. 281–8. Libraire Albert Blanchard, Paris.

Walker, S.C. (1850). Examination of Kirkwood's Analogy. *Am. J. Sci.* **10,** 19–26.

Waterfield, R.L. (1938). *A hundred years of astronomy*, p. 225, Duckworth, London.

Weber, W. (1846). Elektrodynamische Maassbestimungen über ein allgemeines Grundgesetz der elektrischen Wirkung. *Abh. Begründ. k. sächs. Ges. Wiss. (Lpz.)* 211–378 (*Werke*, Vol. 3, pp. 25–214, Springer, Berlin (1893)).

—— (1848). *Annln. Phys. Chem.* **73,** 193–240; transl. by R. Taylor, *Sci. Mem.* **5,** 489–529 (1852) (*Werke*, Vol. 3, p. 215, Springer, Berlin (1893)).

—— Elektrodynamische Maassbestimungen insbesondere über das Princip der Erhaltung der Energie. *Abh. k. sächs. Ges. Wiss. math. phys. Kl. (Lpz.)* **10,** 1–61 (*Werke*, Vol. 4, pp. 247–99, Springer, Berlin (1894)).

—— (1875). Ueber die Bewegungen der Elektricität in Körpern von molekularer Konstitution. *Annln Phys. Chem. (Lpz.)* **156,** 1–61 (*Werke*, Vol. 4, pp. 312–57, Springer, Berlin (1894)).

Weber, W. and Kohlrausch, R. (1857). Elektrodynamische Maassbestimmungen insbesondere Zuruckfahrung der Stromintensitats-Messungen auf mechanisches Maass. *Abh. k. sächs. Ges. Wiss. (Lpz.)* **5,** 265.

Whipple, F.D. (1950). A comet model. I. The acceleration of Comet Encke. *Astrophys. J.* **3,** 375–94.

White, A.D. (1960). *A history of the warfare of science with theology in Christendom*, Vol. 1, p. 154. Dover Publications, New York.

Whitehead A.N. (1922). *The principle of relativity.* Cambridge University Press.

Whiteside, D.T. (1970). The mathematical principles underlying Newton's *Principia Mathematica. J. Hist. Astr.* **1,** 116–38.

—— (1976). Newton's lunar theory: from high hope to disenchantment. *Vistas Astr.* **19,** 317–28.

Whitrow, G.J. and Morduch, G.E. (1960). General Relativity and Lorentz-invariant theories of gravitation. *Nature (Lond.)* **188,** 790–4.

—— (1965). Relativistic theories of gravitation. *Vistas Astr.* **6,** 1–67.

Whittaker, E.T. (1953). *A history of the theories of aether and electricity*, Vol. 2, Chap. 5. Thomas Nelson and Sons, London.

Wien, W. (1900). Ueber die Möglichkeit einer elektromagnetischer Begründung der Mechanik. *Arch. Neerl. Sci. exact. nat., Ser. 3* **5,** 96–107 (volume dedicated to Lorentz on the 25th anniversary of his doctorate).

—— (1901). Considerations on gravitation. *Annln Phys. (Lpz.) Ser. 4* **5,** 501–13.

WILLIAMS, J.G. *et al.* (1976). New test of the equivalence principle from lunar laser ranging. *Phys. Rev. Lett.* **36,** 551–4.

WILKENS, A. (1904). Zur Elektrontheorie. *Vierteljahrsschr. astr. Ges.* **39,** 209–12.

—— (1906). Zur Gravitationstheorie. *Phys. Z.* **7,** 846–50.

WILL, C.M. (1971). Relativistic gravity in the solar system. III. Anisotropy in the Newtonian gravitational constant. *Astrophys. J.* **169,** 141–55.

WOLF, M. (1900). Ueber die Bestimmung der Lage des Zodiakallichtes und des Gegenschein. *Sitzungsber. k. Bayer. Akad. Wiss. Munch.* **30,** 197–207.

WOLTJER, J., JR. (1914). On Seeliger's hypothesis about the anomalies in the motion of the four inner planets. *Proc. k. ned. Akad. Wet., Sect. Sci.* **17,** 23–33.

WRIGHT, A.W. (1874). On the polarization of the zodiacal light. *Am. J. Sci.* **7,** 451–8.

ZENNECK, J. (1903). Gravitation. *Encykl. math. Wiss.* **6** (1), 25–70.

ZÖLLNER, F. (1872). *Ueber die Natur der Cometen*, p. 334. W. Engelmann, Leipzig.

—— (1882). *Erklärung der universellen Gravitation aus den statischen Wirkungen der Elektricität und die allgemeine Bedeutung des Weber'schen Gesetzes, mit Beiträgen von Wilhelm Weber nebst einem Völlstandige Abdruck der Originalabhandlung: 'Sur les forces qui régissent la constitution intérieure des corps aperçu pour servir à la détermination de la cause et des lois de l'action moléculaire' par P.F. Mossotti.* Fock, Leipzig.

INDEX